Becoming Martian

Becoming Martian by Josh Richards
Copyright © 2017 by Josh Richards. All rights reserved.

No part of this book may be reproduced in any written, electronic, recording, or photocopying without written permission of the publisher or author. The exception would be in the case of brief quotations embodied in the critical articles or reviews and pages where permission is specifically granted by the publisher or author.

Although every precaution has been taken to verify the accuracy of the information contained herein, the author and publisher assume no responsibility for any errors or omissions. No liability is assumed for damages that may result from the use of information contained within.

Books may be purchased by contacting the publisher and author at:
www.becomingmartian.com and www.joshrichards.space

Published by LaunchPad Speaking
Editors: Kate Iselin, Shelley Richards, Lisa Stojanovski
Cover Design & Artwork: Adam Peter Scott

ISBN: 978-0-648135-60-9
1. Science 2. Wit/Humor

First Edition
Printed in Australia

Becoming Martian

Josh Richards

Acknowledgements

It's genuinely embarrassing how long I've been talking about writing this book... from the seeds of an idea in late 2012, to an astronaut application in 2013, a jumbled mess of half-written ideas in 2014, the madness of being shortlisted in 2015, thrashing out in 2016 exactly *why* I signed up for a one-way mission to Mars...to finally being here. I may finally have put the words to paper, but what you're really reading here is the product of an entire tribe of people lending some truly incredible support. There is absolutely no way I would have been able to write about how humanity will change in body, mind and soul by colonising Mars without staying with some amazingly generous friends - thank you so much to Chris and Yolanda in Perth, Joanne and Peter in Mudgee, Jess in Hawaii, and Ben and Cariann in LA for giving a ukulele-carrying space nomad shelter to write this.

This book's stunning cover design is thanks to Adam Peter Scott, who has taken many of my half-thought out design ideas and turned them into incredible posters and book covers ever since I started down the path to Mars in 2013. Thank you again Adam - I'm always amazed by your work.

There's also no way this book have been possible without my "Carl Sagan" supporters on Patreon too: James Nelson, Dianne McGrath, Pull2g, Fay Wells, Mary Austin, Ross Hayes, and Megan Papiccio. Your incredible patronage drives me to keep writing every day, to share things I might otherwise hide, and are the reason I can afford to write *and* eat! Thank you for your support, and I can't wait to share the development of my future projects with you just as I have in the writing of this book.

This entire book would still be a barely readable mess without three incredible editors though. I'm eternally grateful for Kate Iselin's talent at re-arranging my sentences into something sensible, for Lisa Stojanovski's relentless fact-checking and astounding knowledge of everything space, and for the proof-reading superpowers of Shelley Richards (Mum). You're three of the most amazing people I know, and I'm forever grateful you each gave so much to turn this jumble of words into something to be proud of.

This book simply wouldn't have been possible without each of you - thank you so much for believing in me.

Martian Body

Sitting on the edge of the couch, mouth agape, I was staring at the most beautiful woman I had ever seen. She smiled softly, floating before me like a flame-haired goddess. Without warning another passenger appeared from the right of the screen, seemingly on a collision course with this perfect being, but with just the slightest push of her finger she sent him spinning away again into the distance. This floating ginger Diana turned back to me, smiled that most glorious of smiles, then effortlessly sailed away out of frame like a dream. Abruptly, the scene changed to a shot of strangers in blue jumpsuits bouncing weightlessly around inside a padded aircraft, with the sounds of angels singing in my head slowly fading back to the overly enthusiastic American narrator describing parabolic flight training... and she was gone.

For weeks I'd been tirelessly working my way through a documentary series on the challenges of sending humans to Mars, and to be honest the eye-candy was generally dismal. No disrespect to the likes of Professor Paul Delaney or Dr Robert Zubrin, but after literally hours of watching aging white men talk to the camera about the finely-tuned personality dynamics required for deep space exploration, I was yet to see much evidence of this "mixed gender crew" everyone was so keen to send to Mars. My initial, primal, "Who are you and will you bear my children?" response to the floating redhead subsided, and as I picked myself up from the puddle I'd formed on the floor, I had a horrible, dawning realisation: if I were ever to actually meet this majestic space unicorn, it would probably be while I was stuck to the floor of an aircraft during a 2g climb, hurling up breakfast into one of those sarcastically labelled "Motion Sickness Discomfort Bags", impotently waving my arms around like a sea turtle stranded on its back while she told me she didn't date other gingers because of the in-flight fire hazard.

You see, weightlessness isn't all champagne, floating red hair, and Strauss's Blue Danube. You might gape slack-jawed at the wondrous freedom of micro-gravity from the comfort of your lounge room, but modern humans have also spent the last 2.8 million years eating, shuffling, and shagging in the consistent pull of Earth's gravity. So while your mind is buzzing at the idea of zero-g backflips, the rest of your body should immediately start screaming "AHHHHHHHH!!! WHY?! Hang on, is that... wait, I think I've got... NOPE - MOTHER OF MONKEY ZEUS, WHAT EVEN IS THIS? WHY CAN I TASTE PURPLE RIGHT NOW?"

At the start of the 1950's Gemini program, NASA wanted its future astronauts to have a tiny taster of what micro-gravity is like. The idea was that they would get a sense of how to move themselves and their equipment around without the binding embrace of gravity while also observing how their bodies reacted to the changing forces. So they ripped all the seats out of a C-131 Samaritan military cargo plane, covered the cabin with white cushions so it looked like a padded white cell with a curved roof, and then started flying this winged roller-coaster through the sky on what was benignly referred to as "parabolic flights".

Just seconds from filling their helmets with carrots and peas
[Credit: NASA]

Each parabola is broken into two parts that are filled with wildly different levels of joy and despair. For the first 90 seconds of the flight, the aircraft climbs at a rather aggressive 45 degrees, at which point you'll be stuck to the floor with nearly twice the force of gravity trying to force your stomach out through your back. But as the aircraft reaches 35,000 feet, the pilot gently arcs the plane out of the climb and straight into a 45 degree dive, so that for about 25 to 30 seconds your body is still moving up while the plane arcs downwards. Done at the right speed, you and your fellow passengers will be weightless, which is great, because now instead of your stomach trying to come out of your back it's lurching forward trying to float in front of you. Delicious. You then go back into a 45 degree climb to do it all again: over a standard two- to three-hour NASA training flight, the aircraft will do 40 to 60 of these parabolas. I'll leave it to you to guess why, sixty years later, astronauts still refer to these flights as the "Vomit Comet".

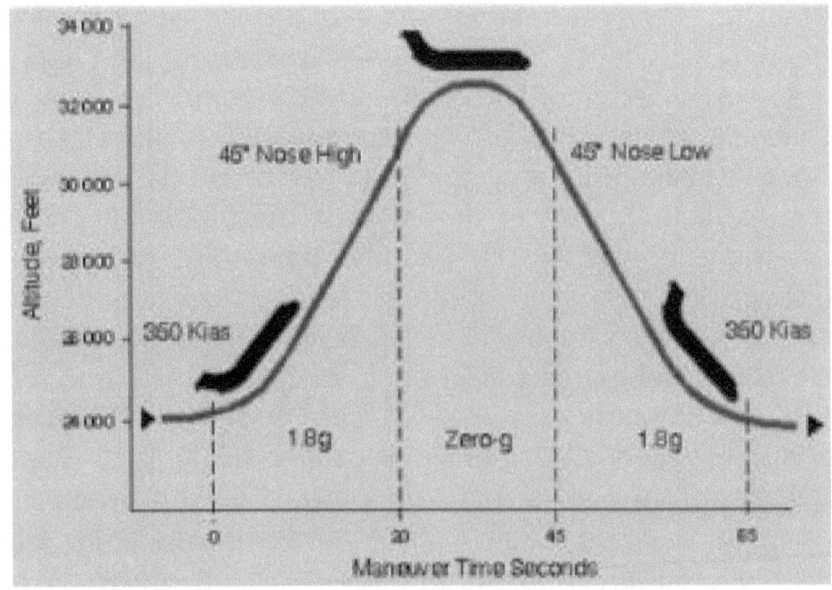

Motion sickness in a deliciously nifty diagram
[Credit: NASA]

In the mid-70s, NASA replaced the original C-131s with two KC-135 Stratotankers that stayed in service until 2004. Much like everything that survived the 80's, NASA even tried slapping on some shoulder pads and skin-tight lycra by renaming the Stratotankers the "Weightless Wonders", but to no effect. The "Vomit Comet" nickname has lived on like the Dread Pirate Roberts of motion sickness. Later, in the nineties, there was even an attempt later to rename the aircraft the "Dream Machines" as part of another sexy rebranding, but unless your idea of a sexy dream resembles a David Lynch-esque nightmare where re-tasting the spaghetti bolognese you had a few hours earlier forms an important part of a bizarre erotic fantasy involving the Log Lady...chances are you're still going to have a bad time no matter what the aircraft is called.

Of course, sexy rebranding isn't a bad thing when it might genuinely reduce passenger fears. According to John Yaniec - lead test director of NASA's Reduced Gravity Program for 15 years - anxiety is the biggest contributor to air sickness among passengers, and the chances of re-visiting lunch seem to follow a rule of thirds: "one third violently ill, the next third moderately ill, and the final third not at all", which also matches up pretty closely to how Ron Howard and the stars of Apollo 13 fared while filming the movie's weightless scenes. Over ten days, 612 parabolas, and four hours of cumulative weightlessness, the scorecard showed Gary Sinise and Kevin Bacon regularly filling their vomit bags, and Tom Hanks and Ron Howard feeling green but managing to keep it all down. But Bill Paxton? He was zooming around, grinning, without a care in the world on every parabola, and I can only hope he was also having flashbacks to playing Private Hudson in Aliens and occasionally screaming, "We're on an express elevator to hell, going down! WOOOO HOOOO!".

So, it's not all airborne despair; nor do you have to be a trainee astronaut or a Hollywood star to experience weightlessness on a parabolic flight. For everyday civilians wanting to get a tiny taste of space, a 90-100 minute flight aboard Zero-G Corporation's *G-Force One* might be as close to the full physiological nightmare of weightlessness as you might be able to get. Founded in 2004 by Peter Diamandis, astronaut Byron Lichtenberg and NASA engineer Ray Cronise, the Zero-G corporation offers regular parabolic flights all over the US for a cool $5000US per person, and thankfully, they also do it with a surprisingly low vomit ratio. It seems most people are okay for about the first 15 parabolas, but then start to go green at around 20, and the cascade hurling is usually in full force by the 25th. So instead of subjecting paying customers to a 3-4 hour flight involving 40-60 parabolas like NASA does to its astronauts, Zero-G avoids the dry-cleaning by only performing 12-15 parabola over a flight. It might only equate to about 5-6 minutes of weightlessness, but a slew of ex-girlfriends will attest this is plenty of time for someone like me to have fun and make an idiot of myself in front of dozens of people we don't know. Unfortunately I'm yet to experience a parabolic flight myself, because if I had I probably wouldn't be writing a book about going to Mars: I'd be sitting on a porch playing banjo and enjoying domestic bliss with my curly-haired ginger wife and our half dozen soulless ginger children.

Medically speaking, the nausea of motion sickness stems from a mismatch between what we're seeing, and what the tiny loops of fluid in our inner ear - the vestibular system - are telling the brain. If your inner ear is saying you're spinning and bouncing around but your eyes say you're not moving (like when you're inside a parabolic aircraft), then your brain thinks you've been poisoned and gets your hurling reflex cranking to rid your body of the poison it assumes you've consumed. Likewise, if your inner ear says you're standing perfectly still but your eyes believe the world has been flipped upside down, you're also probably going to be tasting lunch twice too.

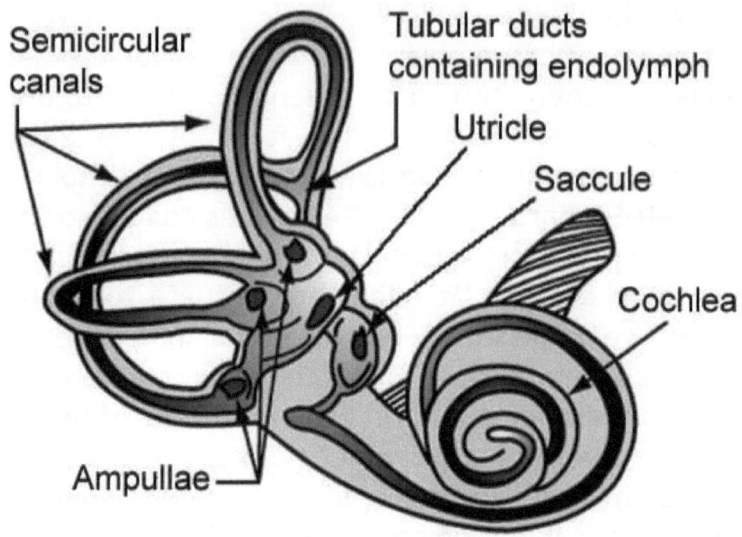

These are in each of your ears telling you which way is up
[Credit: NASA]

The quickest and easiest way to ease the nausea and re-establish some sense to your world is to simply find a window and look out to the horizon. Not only does this give your visual system a fixed frame of reference that will partially subdue the vertigo, it also provides a psychological "horizon" that you can pin your hopes and dreams on. But, as an ex-girlfriend once told me, there's no "horizon" when one of you is going to spend 7 months hurtling through the darkness of interplanetary space on a one-way trip to Mars. With nowhere to look to but the yawning abyss to subdue your motion sickness and relationship issues, the best option is, genuinely, to curl up in a ball and cry yourself to sleep. The tears themselves do very little, but closing your eyes stops the visual element from confusing your brain's balance system, and if you do actually manage to sleep you'll get a few hours bliss to forget about motion sickness and instead dream of giant hammocks, bouncy castles, and emotional security. Much like an emotional, wailing infant, you'll find chewing on things can ease the nausea too. You obviously don't want to eat anything substantial out of fear of adding to the washing machine that has replaced your stomach, but light snacks and chewing gum appear to at least distract nausea sufferers. There's also evidence that ginger can help: chewing ginger root or drinking ginger-infused tea won't stop the raw sensation of nausea, but it's been proven to be an effective herbal remedy to reduce vomiting. Chewing on an actual ginger person, however, will likely result in physical violence by making them "rangry".

Even if you're Bill Paxton, you'll still want to take some sort of medication to ease the trauma of bouncing around inside an airborne roller-coaster. After a few days filming inside the vomit comet for Apollo 13, Tom Hanks got a little too confident and decided to skip his daily dose to see what it would be like unmedicated. This was not a mistake he would repeat, spending most of the flight filling the cabin with his lunch. While there's plenty of remedies that claim to treat motion sickness that are "all natural with no drugs, artificial additives or stimulants" and contain "only the freshest, highest quality chamomile, lavender and frankincense oils", most space agencies like to give their trainee astronauts medication that actually works, instead of simply leaving them smelling like vomit and potpourri. Same goes with those band things that put pressure on "Nei-Kuan" point of your forearms: by all means give it a go, but the scientific consensus is that pharmacology and psychology are more likely to win the nausea battle.

By far the most commonly prescribed motion sickness medication is dimenhydrinate, commonly sold as

Dramamine. Combining a nausea-quelling antihistamine with a stimulant not dissimilar to caffeine, Dramamine will help reduce the nausea associated with motion sickness... but it might also knock you out in the process. While other medications such as meclizine may not put you in the land of nod quite as quickly, all modern motion sickness medications make people at least a little bit drowsy because they work by telling your central nervous system to calm down instead of freaking out and bringing up breakfast, which is why most aviation authorities worldwide prohibit pilots in command from using motion sickness medication at all, and why the warnings on the boxes recommend not to take it and operate heavy machinery. Warnings that I'm guessing probably also apply to flying a multi-billion dollar spaceship to Mars...

There's also the minor issue that when these drugs start to mess with your central nervous system they can also make you trip harder than John Lennon writing Yellow Submarine. In sufficient doses, Dramamine acts as a deliriant, with recreational users talking about "Dramatizing" or "going dime a dozen", and giving the drug a whole series of different street names like "dime", "D-Q" and "drams"...all of which I just pulled straight off Wikipedia because I have no experience with Dramamine-induced delirium whatsoever. But my Mum does! A few years ago my parents went on a scuba diving trip out to the the Rowley Shoals, a series of atolls about 260 kilometers out from Broome on the north-west coast of Australia. While Dad has always prided himself on his cast-iron stomach, the eight hour boat trip to the shoals took its toll on Mum. Luckily there were some friendly Germans on the boat too, and rather than indulging in their national past time of schadenfreude by laughing at her suffering, they gave her a couple of tablets that they assured would help the nausea... and it worked! Mum didn't feel an ounce of nausea while she chased non-existent "molecules" around the deck of the boat for the next few hours, trying to scoop them up gently in her hands and showing them to everyone on board. So the Germans had their schadenfreude after all, only with less "projectile vomiting" and more "Australian mother of two hilariously tripping her face off while hundreds of kilometres out in the Indian Ocean during heavy seas".

While Dramamine might be the solution for parabolic flights and regular car/seasickness, the best option for astronauts seems to be the far stronger and longer lasting scopolamine. Usually coming in the form of a VERY sexy (read: not sexy at all) transdermal patch that gets stuck behind your ear like a leech, scopolamine patches slowly administer the drug over several days and provide astronauts with nausea relief during their initial adaptation to life in space. Just make sure you wash your hands if you touch the patch, though, as it'll cause blurred vision if you manage to get it in your eyes. Scopolamine still causes drowsiness, so the military found a solution for their fighter pilots: "Scop-Dex", or scopolamine mixed with dextroamphetamines. That's right: the air force took heavy-duty motion sickness medication and mixed it with the pills your friends used to buy/steal from the ADHD kid in high school before trying dancing to Moby. Scientists didn't believe it was even possible to dance to Moby, but the kids you went to school with proved it while the ADHD kid just bounced awkwardly in the corner as the unmedicated control sample.

Space agencies are obviously keen to avoid having astronauts vomit on expensive control panels, doze off while flying a spaceship, and throw out all the supplies to make room for an all-night space rave. As a result, a huge amount of research is continuing into how nausea from motion sickness can be minimised in space without medication. One of the most promising technologies is the use of strobe lighting and LCD shutter glasses that flicker at a high frequency so as not to interfere with your vision. Initial experiments with participants on the ground and during parabolic flights have now shown that a short-duration flash - four to eight times per second - significantly reduces the symptoms of motion sickness. So while I might not be drowsy or vomiting into a paper bag when I finally meet that ginger sky unicorn on a parabolic flight, I'll probably be suffering the indignity of having to wear NASA-designed shutter shades and feeling like I've helped film a Kanye West video in space.

Your aunties in the world's worst Daft Punk tribute band
[Credit: New Scientist]

Speaking of indignities, if you were hypothetically to type "zero g corporation redhead" into google image search, Jake Gyllenhaal is the 8th picture you'd see. Probably. When you eventually found your ginger space unicorn on the 14th page of results, it'd also be instantly obvious she's not really a redhead, and all your ginger militia-founding hopes instantly disintegrate right there. In retrospect, though, if I'm falling in love with a women based on about eight seconds of footage from a documentary series made in the late 90s, I'm probably not in the right place emotionally to be contributing to the gene pool anyway.

For all the wonder and beauty of space, all the spiritual awakening and inspiration that astronauts report after seeing our beautiful, fragile planet from a perspective that doesn't show borders or racial and religious differences, just one Earth... chances are you're still going to be tasting your own stomach acid. Your life-altering spiritual experience is being tainted by a little thing space medicine experts casually refer to as "S.A.S." or Space Adaptation Syndrome. And we can't talk about Space Adaptation Syndrome without talking about Senator Jake Garn and Margaret Thatcher...

<u>Tasting Purple With Your Ears</u>
In 1985 US Senator Jake Garn flew aboard space shuttle Discovery's STS 51-D mission as part of a "Senate Oversight" trip, aka "Getting to fly in space because you're a US senator in charge of the Senate Appropriations Committee". Although Senator Garn had US taxpayers cover his joy ride into space, he paid for it once he was there: for most of his week in space he was practically incapacitated with nausea. You'd think that after the initial lurch of being fired into space, a few minutes of majestically floating through the cabin in microgravity would quell any

motion sickness you might have felt before... but the brain can be a cruel and hilarious thing.

While logically you might process which way is up or down - possibly aided by the massive signs stuck everywhere saying "Port", "Starboard", "Up" and "Down" - we humans haven't yet evolved to the point where our inner ear can read. So while your eyes are telling you, "That way to the cockpit!", once again the fluids in your inner ear canals are swirling randomly screaming "WHERE IS THAT TINY BOX OF DESPAIR THEY CALL A TOILET?". (Which is where Senator Garn spent most of his week in space.)

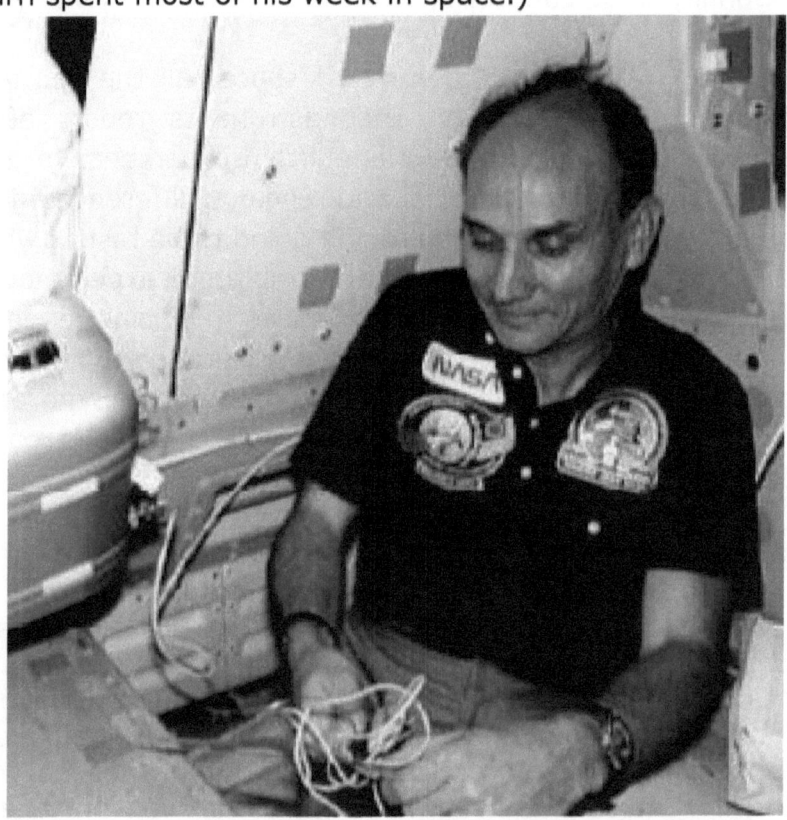

Jake Garn in a rare moment when food was *inside* his body
[Credit: NASA]

While officially there is no scale for measuring just how affected by space adaptation sickness an astronaut is, according to Shuttle astronaut Mike Mullane there is an unofficial "Garn Scale" among NASA astronauts: 0 on the Garn scale means no motion sickness, and "1 Garn" being completely and utterly incapacitated. Most trained astronauts don't get above 0.2 on the Garn scale, experiencing only a little discomfort for the first few days as their body adjusts to life in microgravity, but Senator Garn never fully adjusted during his week in space. NASA did name their primary training facility for shuttle and space station astronauts the "Jake Garn Simulator and Training Facility" after him though, so I can only assume the experience was worth it.

Mixed in with all this vomit-inducing vertigo is the fact that humans instinctively identify each other as having two eyes above a mouth. This evolutionary shortcut remained relatively unchallenged for nearly 3 million years of human-primate evolution until people saw Margaret Thatcher's face; And before you conservatives get all huffy about what sounds like an unexpected insult to the Iron Lady, trust me when I say it's an *evidence-based* insult because Margaret Thatcher's face literally redefined cognitive science.

In 1980, psychologist Peter Thomson at the University of York showed that we can only effectively interpret someone's expression when their eyes are positioned above their mouth, and that it's no easier to read someone's expression if their face appears inverted but the eyes and mouth have been manipulated to appear upright. To prove his point, Professor Thomson then horrifically demonstrated it by inverting Margaret Thatcher's eyes and mouth in the picture below:

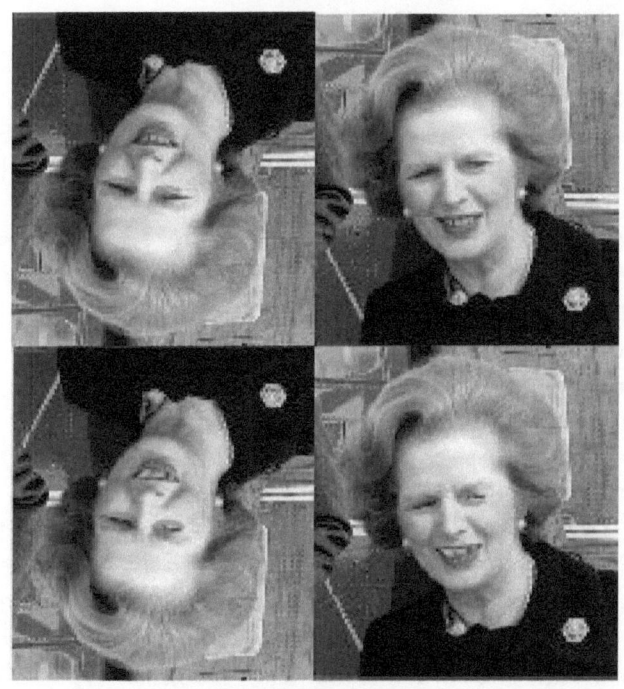

She doesn't actually look *that* different, but if I had to see this then so do you
[Credit: Public Domain]

Thus the "Thatcher Effect" was coined. Without a fixed "up" or "down" in space, astronauts not only struggle to understand expressions when they're looking at someone who is upside-down, but they can also fail to identify who they're looking at altogether. Imagine being on a seven-month trip in a cramped capsule on the way to Mars, working away on a flight path trajectory to make sure you don't skim off the Martian atmosphere into the darkness of space when suddenly the one crew member you're not sleeping with floats over with a big stupid look on his face... and plants a juicy kiss on your lips. While you might be excited and a little nervous at the prospect of this unexpected turn of events, there's a really good chance they've just mixed you up with someone else because you were upside down. Although I'm guessing most people look rather pretty when you've been locked in a tin can for seven months...

Space agencies have a simple but ingenious way to avoid all of this partner-swapping confusion: put different crews in different-coloured flight suits. Because of the rolling rotation of crews on the International Space Station, crews can be on board with up to six other astronauts. They might know each other well after years of training together, but that doesn't mean their eyes won't play tricks on them in microgravity. By simply putting each crew in their own colour-patterned flight suits, they can avoid hilarious partner-swapping mix-ups.

Expedition 37 and 38 with great uniform diversity, not much gender diversity though
[Credit: NASA]

A crew heading to Mars will have spent close to a decade training together, so hopefully there won't be too many unexpected shenanigans; plus the surface of Mars has roughly 38% of Earth's gravity to provide a clear "up" and "down" to stop things getting topsy-turvy. That reduced gravity will also make it easier to hang from the shower head if you want to replicate the upside-down "Kiss In the Rain" from Spiderman too, so don't listen to ground control on Earth telling you what the behavioural guidelines and water restrictions are when you and the Crew Commander are pretending to be Kirsten Dunst and Tobey Maguire. Go get 'em, Tiger.

Bigger, Higher, Bustier, Puffier
While your stomach, eyes, and inner ear are still adjusting to the experience of not having an up or down, the rest of your body starts to freak out about the rest of the space experience. After literally decades of the constant pull of gravity bearing down on you every second, suddenly you're free to float and expand in every direction like Uncle Frank did after he retired. But since our bodies evolved with that constant gravitational pull, absolutely everything starts to go haywire. Earth's gravity naturally draws the fluids in our bodies downward, so our heart is designed to keep the blood flowing up to our chest, arms and brain. Without that gravity to work against, though, all the fluids in our legs shift upwards, which turns out to be *great* for your body image! Wrinkles fill out to make everyone look younger, arms swell to look stronger, breasts swell to look perkier: it's basically "Baywatch" but with less sand and none of David Hasselhoff's chest hair.

Discontent with this sudden return to full and buxom youth, your jerk of a brain immediately decides to ruin *everything*. While you're consciously thinking, "Hey this space business is pretty good for my pecs!", 3 million years of evolution kicks in and your primate brain says, "Oh, *this* isn't normal! What's with all this extra fluid making you look a little bit sexy? Let's just go ahead and get rid of that...", which is when you start peeing out all that "excess" fluid your brain thinks you're carrying and within a few days all your squishy bits have shrivelled up. So while you'll look ten years younger when you first get to space, evolution quickly snatches those naive hopes away as you transform from Duffman into Mr Burns.

A New Hope (For Shorties)

While your upper body is puffing then shrivelling, the muscles in your spine decide to get in on the action too. While most of us have spent the majority of our lives standing up, the sudden freedom from gravity's downward pull causes the spine to relax as well, with astronauts growing several centimeters during their first few days in orbit. That sounds great if you're vertically-challenged like me, but when you actually stop and think about it you'll start to realise just how much pain that's going to cause. It's not uncommon for some astronauts to be nearly incapacitated by back pain during their first few days in space as their spine adjusts. All this expansion also puts astronauts at a higher risk of suffering a spinal disc herniation - the dreaded "slipped disc" - where the squishy, shock-absorbing part inside an invertebrate disc is pushed out through a tear in the disc's outer layer and pinched between the vertebra.

Being a "fun-sized" astronaut is definitely an advantage in space, even if on Earth it means you have to gaze up at attractive women and discover that "I'm an astronaut candidate training for a mission to Mars" only helps you pick up a startling sense of self-loathing. With a shorter spine and less distance between our head and feet, the smaller among us can endure g-forces more efficiently than others with average or giraffe-like stature, because there's simply less room for your blood to run away to as it's being pulled from your brain.

Unsurprisingly, a shorter crew also means more room to move around inside what are often claustrophobic spacecraft. Paolo Nespoli flew to the ISS in 2010 alongside Cady Coleman and Dmitri Kondratyev as part of Expedition 27. While Cady had the rather adorable issue of being so short she struggled to reach switches from her Soyuz seat, and Dmitri's biggest challenge was smiling, Paolo was physically too tall to fly in a standard configuration capsule. Towering at 188 centimeters, the only way they could physically fit him in was to custom build his seat, shifting it a couple of degrees off the optimum angle and subsequently increasing the risk he'd break his back when the capsule landed back on Earth. As with everything on this wonderfully litigious planet, any time you want to do something that's even fractionally *more* dangerous than flying in space, you have to sign a waiver. So while Paolo spent six months in an orbital space station, orbiting the earth at 27,600km/hr in an environment where a single micrometeorite breaching the hull could have killed everyone aboard in seconds... he still had to sign away his right to sue if he injured his back MORE than the 9g impact already would.

Paolo Nespoli stretching out on the Space Station
[Credit: NASA]

Paolo Nespoli remains the tallest astronaut to ever fly in a Soyuz, so it's probably best not to send clones of Andre the Giant to space. Strap a bunch of Oompa Loompas into a capsule on the way to Mars though, and they'll happily bounce around inside with room to spare, making chocolate and singing cheerful songs about murdering obnoxious children before the capsule door is even closed.

The Floating Dead
At some point the buzz of *being in space* is going to start to wear off, and you'll want to try and sleep. When you do decide to stop staring at the stars, you'd probably expect it to be literally like snoozing on Cloud 9 - a blissful affair where you float weightlessly, breathing unhindered by gravity acting on your lungs, silently dreaming in the perfect darkness of space. Nothing could be further from the truth.

For starters, space is noisy. Really noisy. During the construction of the International Space Station, astronauts regularly recorded noise levels over 70 decibels (dB), as well as spikes up to 90 decibel during sleeping periods. That's like having a really loud alarm clock at 70dB going off all the time, that wailing banshee of a child in the local supermarket screaming about his favorite cereal fairly regularly at about 80 dB, and occasionally having someone throw the alarm clock into a blender at about 90 dB while you sleep. What's making all the noise? Pumps and fans. Heavy-duty ammonia pumps are needed to help cool the station, motors need to constantly turn the station's solar panels to face the Sun, fans need to cool the computers monitoring the life support... there's a whole raft of things up there making noise. It's not radically different to the constant hum you get from the engines on a commercial jet, except in space there's no external atmosphere to damp down the noise or vibrations. All of those thumps and whirs just rattle through the entire structure, like being on the bottom bunk of a bunk bed in a hostel while the guy above you gets laid every night.

Even with a pair of heavy-duty earplugs, sleeping in space still isn't easy. Nothing in space stays where you left it for long. If you let go of a pen it'll float there for a few seconds, but turn around and it's gone, drifting with the spacecraft as it orbits, or carried away on some imperceptible current of air. That's why there's velcro on everything from pens to spoons: so things stay where you stick them. Likewise if *you* are not strapped into something when drifting off to sleep, you might wake up to find you've drifted straight into the toilet. People think sleepwalking is terrifying, but imagine startling yourself awake as you casually bump into the airlock depressurization handle. In space no one can hear you scream, but inside the spacecraft your squeals of terror are going to scare the living Monkey Zeus out of everyone else on board.

At least if you *were* bouncing around the station in your sleep you'd be getting a little fresh air. On Earth our exhaled breath moves upwards because it's warmer than the surrounding air, but yet again there is no "up" in microgravity for the exhaled breath to move too. Without a vent or fan to circulate your exhaled breath, a bubble of carbon dioxide can build up around a sleeping astronaut's head to the point they become starved of oxygen, or at the very least wake up with a splitting headache from carbon dioxide poisoning. That's if you're not already waking up screaming because your inner ear is giving you those horrifying dreams where you think you're falling. While you can clip a sleeping bag anywhere on the station and zip yourself up inside, crews tend to use the well ventilated, telephone-booth sized sleeping cabins that provide a modicum of sound-proofing and personal space in order to get away from the noise of the station and their own scream-filled dreams. Besides giving astronauts a little bit of room for personal items, the cabins also have a sleeping bag strapped to one wall that keeps the astronaut in one place. These prevent crew members from bouncing around inside their cabin as they sleep, as well as giving their head something to "lay" on to reduce that horrifying sensation of falling as they doze off.

Now it might come as a shock to you, but none of this is exactly ideal when you're asking astronauts to operate an orbital laboratory worth $150 billion US that's flying 400 kilometers above the Earth at over 27,600 kilometers an hour. You're puffy, sleep deprived, probably making out with your coworkers, and you've got a screaming headache. Although, this is still a lot better than what some people initially thought being in space would be like.

Petticoat Pandamonium

The invention of the steam locomotive in the early 1800s brought both the excitement of innovation, and long-distance travel. Unfortunately, because humans are idiots, it also brought an overwhelming fear from the medical community that any man moving faster than 30 kilometers per hour would suffocate, and a woman would simply disintegrate from the mind-bending idea of exceeding carpark speed limit. There were even suggestions that railways would need to be shielded from view by brick walls, because surely anyone seeing a train travel that fast would lose their mind and cows would be so frightened their milk would curdle.

Whhhhy hoooomaans soooo stooopid?
[Credit: Public Domain]

Unsurprisingly, there were also some pretty ludicrous predictions made about what could happen to someone in a weightless environment before Yuri Gagarin became the first human in space on April 12th 1961. Doctors had countless theories that an astronaut's heart would simply stop beating, that they wouldn't be able to breathe or swallow food, and that they'd even instantly go blind as soon as they became weightless. Trials in parabolic flights showed that most of this was completely unfounded, and as the Soviets and the US started regularly launching people into space everyone stopped being so hysterical about microgravity. When crews started going into space for several months at a time though, flight surgeons soon realised that being in space really *could* send you blind... and began completely freaking out again.

Weird-Eye For The Space Guy
Halfway through his 6 month stint aboard the International Space Station, John Phillips knew there was something really wrong with his eyes. While the former US naval aviator had proudly proven his 20/20 vision before launch, looking out the station's cupola the Earth looked blurry and out of focus. It wouldn't be until he returned to Earth another three months later that tests would show his vision had dropped to 20/100 - making him almost legally blind - and an ultrasound would show both of his retina had unexpectedly flattened. When Canadian astronaut Dr Bob Thirsk reported needing help to focus cameras because he'd become far-sighted within just weeks of boarding the space station in 2007, NASA knew they had a serious issue on their hands.

Regular in-flight testing of astronaut's eyes using hand-held ultrasound units began during Dr Thirsk's flight, and what became clear shortly afterwards is that around 60% of astronauts spending between two and six months on the International Space Station experience some form of vision degradation.

Dr Bob Thirsk testing his eyes on the Space Station
[Credit: NASA]

What isn't clear, though, is what causes it. Researchers have suggested astronauts are getting too much sodium from their food, too much carbon dioxide from poor ventilation when they sleep, and even that they're exercising too much. The leading theory, however, is that just as fluid shifts upwards when you first arrive in space to give you a puffy face, spinal fluid also shifts up and increases the pressure in the brain cavity; except instead of losing that spinal fluid at the same time your puffy face disappears, it stays there increasing the pressure in your skull, inflaming the optic nerves which push against the back of your retinas and flattening them while also increasing the pressure inside your eyeballs themselves. As a temporary solution, adjustable focus glasses are kept on the space station, and although it's still a cause for major concern, I guess if everything is blurry at least you'll have a better excuse for making out with the wrong upside-down crew member.

Without gravity to work against when you're moving around, it also doesn't take your douchebag brain long to realise the heart isn't being used as intensely as it would be on Earth, and immediately puts your heart on casual working hours. Within weeks of getting to space your heart starts to shrink and get rounder, turning into that barely-employed acquaintance your girlfriend assured you was, "Only staying for a few days so he can get back on his feet". As the months roll on it gets progressively weaker and fatter, stops looking for work, and eventually turns into this pathetically soft thing that never seems to leave your couch but still manages to find the energy to ask for Doritos, Mountain Dew, and money. Any time you ask it to do a little bit of work to help around the house, your heart spins around and starts yelling, "Woah woah woah - you want me to do WHAT?! I already looked at some jobs sites today, and now I'm a bit tired. Besides, Brain said there's nothing going on, so I can just hang out and watch Oprah". Just like your cheese-fingered house guest, the longer you stay in microgravity, the worse things get: after six months on the space station, scans show that most astronauts have hearts that look like they've won the space-sloth lottery.

Peeing Out Dem Bones

The biggest kicker of life in microgravity - in case not being able to sleep, having your heart go on vacation, or going blind isn't bad enough for you - is what all that graceful floating around does to your bones. Without the regular stress of lifting, moving, and shifting in Earth's gravity, your bones decide to take a holiday along with your heart. All the fluid you're peeing out in the first few days? That's your body beginning to break down your bones because they're not being loaded up by gravity any more. Astronauts are often at a serious risk of developing kidney stones in space because their bodies are dumping the calcium from their bones into their blood that the kidneys then try to filter out. Awesome.

Over a six month expedition, astronauts can lose up to 20% of their bone density, and can require as much time in rehabilitation as they spent in space. So you might survive the entire journey to Mars - spending seven months in a spacecraft doing 40,000 kilometers an hour across 56 million kilometers of dark and unforgiving deep space - survive the truly terrifying landing, and climb down the ladder onto that cold and pitiless planet, only to have your historic first words be: "Dammit Dale, did you seriously just trip and break your hip? WE'RE A BLOODY LONG WAY FROM A HOSPITAL, DALE!".

Both US and Russian flight surgeons have tested all sorts of weird and wacky things to try to stop astronauts' bones from breaking down; from drug cocktails to weird vacuum-pressure pants and even body-squeezing underwear designed to simulate the same pressure as Earth's gravity through your shoulders and feet. There's also been no shortage of folks suggesting huge spacecraft with "2001: A Space Odyssey"-style artificial gravity rings, or even much smaller high-speed centrifuges that astronauts are somehow supposed to sleep in without throwing up. Yet after all the decades of research and study, it seems like good, old-fashioned exercise works better than anything else. Through a research partnership between NASA and the Japanese space agency (JAXA), it's been shown astronauts on the space station who exercise for an average of 15 hours a week - around 2.5 hours a day, six days a week - will experience *almost* no bone density loss. By supplementing exercise with a weekly dose of alendronate - a commonly-used medication for osteoporosis patients - bone density loss stops altogether, along with the risk of kidney stones.

Valeri Polyakov didn't have any fancy osteoporosis medication to help him during his 14 months in space though: he just exercised like hell. To date, he still holds the record for the longest single spaceflight, spending a whopping 437 days aboard the Soviet space station Mir from 1994 to 1995. It was all part of a physiological and psychological simulation of the type of long-duration mission required to get to Mars and back, and by all accounts he proved in the mid 90's that humans were capable of physically surviving space for as long as it would take to get to Mars.

Valeri Polyakov still grinning after over a year on Mir
[Credit: NASA]

Normally after landing in a Soyuz capsule the crew are lifted out by the ground crew and then physically carried to a lawn chair nearby for their initial post-landing medical checks. This is partly because the crew may be struggling to readjust to Earth's gravity, partly to avoid someone being injured by a fall, but mostly because the Russians just like carrying around people who've just come from space. Polyakov refused all of this, climbing out of the capsule himself and walking unassisted to the waiting chair to prove that after such an extended stay in space someone would still be strong enough to walk after a landing on Mars without help from the local Martian population. His first words out of the capsule were "We can fly to Mars!", while an American astronaut witnessing the recovery apparently proclaimed Polyakov looked "Strong enough to wrestle a bear". Although as everyone knows: in Russia, bear wrestle you.

While Polyakov's 14-month mission stands as a testament to human adaptability, it still falls well short of the 3 year missions that government space agencies like NASA are planning for. Concerns about bone density take a rather interesting turn when you start thinking about *one-way* trips, though. Mars One colonists would need to maintain their bone strength during the six- and eight-month journey to survive the higher g-forces during landing, but once they're on the surface they'd be spending the rest of their lives in 38% gravity. My 65 kilograms of raw ginger chaos on Earth would be just 25 on Mars, so naturally my body would try to adapt to that over time. My muscles might lose *some* of their strength, my heart would likely become *slightly* weaker, and I might thankfully be a *fraction* taller; but for the most part we'd be like superheroes: lifting nearly three times more than we could on Earth, running with nearly a third less effort, jumping nearly three times higher… just imagine the weird sex stuff you'd be able to do! It'd be enough to make everyone watching you back on Earth sick.

More Than Man Flu
That's something else you'll really want to avoid in space: getting sick. Today, crews heading to the International Space Station spend three weeks in hard quarantine before launch, locked away in the soviet-era accommodation at Baikonur, with no physical contact with others and the only connection to loved ones and the outside world being through thick glass. Two weeks is roughly the incubation

period for many diseases, so most signs of a possible infection will have developed at least a week before the end of the quarantine. This allows plenty of time for the prime crew to be replaced by a backup crew who have been quarantined over the same period. Provided no one suddenly comes down with Ebola, the Black Death, or even a head cold, then the crew are put through one last indignity before suiting up and climbing inside their Soyuz capsule: their whole bodies are wiped down with alcohol to disinfect anything on their skin, because the absolute last thing you want is some kind of germ or virus on your spaceship.

Not only does microgravity suppress white blood cell production in astronauts by up to 10% during a six month mission (so you've got less chance of fighting an infection off), but experiments on the space station growing wonderful things like E. Coli strongly suggest that bacteria *love* being in microgravity, permanently mutating to give you the space squirts. Spacecraft are built in high-purity clean rooms to protect the sensitive onboard electronics, but it's also to make sure that nothing from someone's truck-stop sandwich gets into the spacecraft and starts growing when it gets to space, turning into a horrifying superbug that stalks and murders the entire crew in their sleep. The small, cramped living quarters and recycled air in spacecraft make the spread of any pathogen inevitable, so even relatively minor infections are dangerous: just getting a head cold can prove deadly. When Commander Wally Schirra developed a cold in Low Earth Orbit during Apollo 7's 11 day test flight, there was no gravity to pull the congestion down from his sinus so the pressure just kept building. Soon the other two crew members, Donn Eisele and Walter Cunningham, were infected too, and the entire crew started to lash out at Ground Control over the radio. There was a mini "mutiny" on board with Schirra refusing to turn on TV cameras because they were "busy" and telling NASA's flight Director Chris Kraft to "go to hell", while the entire crew complained about eating food they'd personally picked before launch. More seriously, they also refused to wear their helmets during reentry, fearing if they couldn't squeeze their nose to equalise the changing cabin pressure, then their congested sinuses could lead to their eardrums bursting. Ground control warned the Apollo 7 crew there would be consequences for refusing to wear the helmets before landing, and sure enough, none of them ever flew in space again.

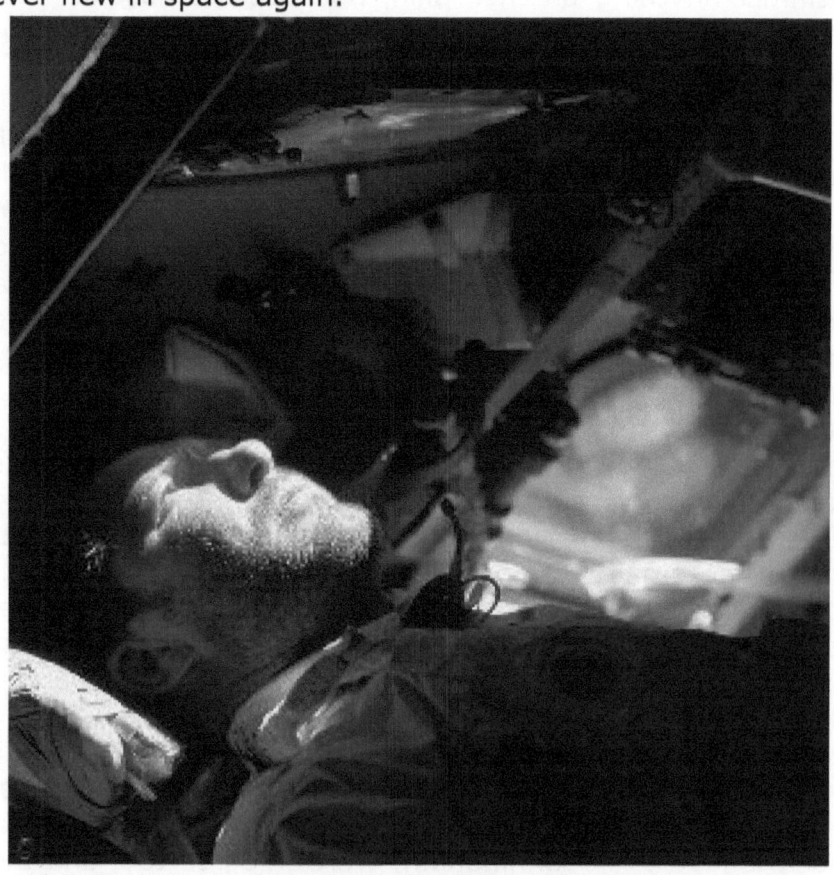

Wally Schirra looking out the window of Apollo 7, giving zero shits about protocol
[Credit: NASA]

While the International Space Station is stocked with antibiotics and flu medication, in an emergency an astronaut can be medically evacuated in less than four hours. But not having a quick way to get to a hospital - like if you're living permanently on Mars - means that getting sick in deep space can easily prove fatal. When an oxygen tank inside Apollo 13 exploded on its way to the Moon, the crew had to cut their food and water to just a third of the minimum in order to survive the next 5 days in the crippled spacecraft. Lunar Module Pilot Fred Haise first reported feeling unwell before the capsule even got to the moon, but dehydration turned what started as a minor bladder infection into a full blown medical emergency. By the time Apollo 13 managed to slingshot around the Moon and land back on Earth, he was practically incapacitated with pain.

Astronauts on the way to Mars won't have the luxury of a relatively quick five-day trip back to Earth - once you're on the way there's no way to just "turn around", so you have to go all the way there before you can head back. Even if you *did* slingshot around Mars and come straight back, the planets would be so far out of alignment it'd take nearly *thirteen months* to get back to Earth, and that's only if you're willing to absorb a huge radiation dose as you dip perilously close to the Sun in the process. For anyone planning to go to Mars one-way the spacecraft simply won't be carrying enough fuel, water or food to last beyond the seven months to get there, much less an extra thirteen months of supplies if something went wrong.

Prevention is definitely better than cure, so to make sure they're not carrying some sneaky strain of zombie virus, astronauts and cosmonauts go through extensive medical testing throughout their entire careers to identify even the most minor things that could become full-blown medical emergencies in space. Even body parts that you might only get checked occasionally can easily cause problems when you're living in any kind of isolated environment like Antarctica or on a submarine, let alone Mars. What happens if someone's appendix bursts on the way there? One option for future Martian colonists might be to have preventative surgery before launch, like having your appendix removed *before* it has a chance to explode in space. Another rather brutal concept might be to use surgery to reduce a colonist's cancer risk by removing the highest risk reproductive areas: double mastectomies for the ladies and testicular removal for the guys. It might be tough to think about, but reproductive tissue is especially prone to mutation so having less of it reduces the risk of developing cancer from ionising radiation. Although, the risks of this sort of preventative surgery mean it's not yet clear if it would even be worth it. The appendix plays an important role in the human immune system and removing it could place someone at even higher risk of infection than they might have of developing appendicitis. Likewise, removing large areas of reproductive material might change hormone levels so much that overall cancer risk actually *increases* instead, so don't get excited and start chopping off any of your bits *just* yet.

One area where the benefits of preventative surgery are clearer is when it involves getting all those pesky wisdom teeth out *before* they burst through your gums and bleed on Mars. Dentistry is something you might not expect an astronaut would have to learn. First aid? Yes. Emergency surgery? Probably. But telling a fellow colonist to say "Ahhh" and asking them about their weekend while you poke around in their mouth is something every future Martian colonist will have to do. We'll need to know how to polish and pull each other's teeth out because there's no dentists on Mars unless we send them there; although scientists in the UK have developed a technique that might make keeping your chompers strong on Mars a whole lot easier. King's College researchers encouraged the bodies of test subjects to redirect tooth-building minerals back into existing teeth using tiny electrical jolts. This "Electrically Accelerated and Enhanced Remineralisation" repairs enamel and keeps teeth healthy, and may mean the regular scrapes and cleans from the dentist can be replaced on Mars with some bumbling ginger idiot jamming a bunch of electrically-charged probes into your mouth instead. But although the idea of me poking around in your mouth is pretty horrendous, it's not even close to the grossness that happening at the other end of your digestive tract.

Unidentified Feculent Objects

Much of my time now is spent visiting schools around the world to talk about colonising Mars and the future of human space exploration, and I'm genuinely grateful for it. It's inspiring to see the eyes of kids light up as they contemplate the wonders and challenges that space presents our species, envisioning a future where *everyone* will be able to explore the solar system regardless of which patch of Earth they were born on or claim loyalty to.

But while these kids envision a future where we explore space together as a species rather than as competing nations, it's guaranteed some other kid in the audience will put their hand up to ask me about how to shit in space. Every time. Not just most of the time. EVERY. TIME. Of course they don't say it that way - it's always, "How do you go to the toilet in space?" - but a quick Google image search (which I know they've already done) proves there's a variety of zero-g hose systems for both male and female astronauts to urinate into. So what these kids are really asking is "How do you shit in space?". It's an understandable question to ask, though: you're in space, you float, other things float too...how does that all work? The short answer is: poorly.

The earliest solution used by the first Soviet cosmonauts and US astronauts were simply nappies. Great big adult diapers like your incontinent grandfather uses. In many ways going to the toilet in space has radically changed, but the nappies have been there right from the start and are likely to remain there for a long time yet. Even today, astronauts conducting a spacewalk on the space station have to wear a nappy - there's simply no chance to bolt inside to use the smallest room on the spaceship when you're buttoned up in a 145 kilo spacesuit, so you just go in your nappy like a month-old baby. If an astronaut ever arrogantly thought they were "on top of the world" during a spacewalk - looking down on us mere mortals from hundreds of kilometers above - then I'm sure having to shit in their spacesuit would be a humbling enough experience to remind them that they're still human.

As humans started to stay longer than a few hours in space, other solutions to soiling ourselves needed to be found. As the kids doing the Google Image search would have found, dealing with liquid waste is relatively simple. With the exception of Valentina Tereshkova - the first woman in space, who wore a nappy just as Yuri Gagarin had to - all the early astronauts and cosmonauts were men, so they just had to connect themselves up to a tube, pee into it, and then vent it into space. The urine would then instantly freeze into tiny droplets that in sunlight would form stunning rainbows all around the capsule, and US astronauts would photograph it to show people back on Earth the "Constellation Urion". Don't you judge me, it's Wally Schirra's joke. I'm cringing at it as much as you are.

Thankfully everyone agreed that trying to make space rainbows with turds was a bad idea, so until the space toilet was developed in the 1980's space explorers had only one option for the solid waste - crapping into a plastic bag and storing it all the way home. Long-duration cave explorers use the same technique to reduce their impact on the pristine underground environment, but the early astronauts and cosmonauts only did it because it was a sticky half-a-step up from a nappy. Of course, the true mechanics of doing this in space become fairly horrifying when you realise nothing falls down the way it should, so it's pretty difficult to make a clean "break-away": it just floats there, so even taping a plastic bag onto your ass isn't always going to stop those space turds escaping to roam free around the capsule.

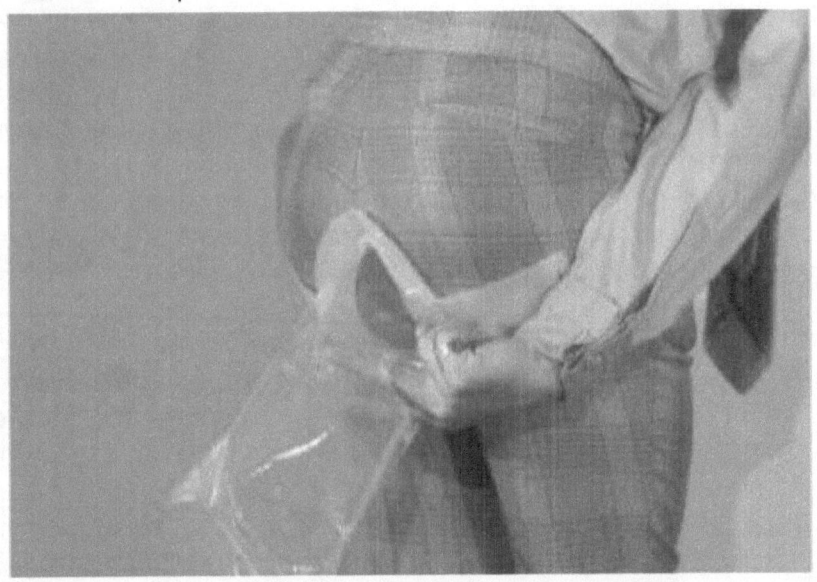

Finding the credit for this image was a nightmare - thanks NASA
[Credit: NASA]

The crew of Apollo 10 were repeatedly terrorised by free-floaters as they orbited the Moon, with NASA's official declassified radio transcripts of the mission including classic comedy lines like, "I didn't do it. It ain't one of mine", "Mine was stickier than that," and, "I don't know whose that is. I can neither claim it nor disclaim it (laughter)".

Besides rogue space turds, the other issue with using a plastic bag is that solid waste doesn't just stay as an unchanging lump - it evolves. Pretty soon it starts to breakdown and release gases, so it's only going to stay *in* the plastic bag until the gas pressure builds up enough for it to burst. Merry Christmas. To avoid exploding crap bags in their spaceship, astronauts used a fantastic powdered chemical compound that killed off the gas-producing bacteria in the waste. Problem solved, right? Kind of. Getting the anti-bacterial powder into the bag in microgravity was enough of a challenge, but once it was there, astronauts still had to squish the bags by hand to ensure it was properly mixed together. There's stories of astronauts during this time being too busy to mix the compound through themselves and having to pass the task off to a crewmate to do the honours. Because what are friends for if you can't squeeze anti-bacterial powder through each other's shit?

With mixed-gender crews, the designers of the space shuttle quickly realised they had a new toilet issue to contend with. Not only were the astronauts over the idea of crapping into plastic bags, they were pretty sick of having no privacy when they had to do it, too. It took women joining the astronaut corps before the idea of an actual space toilet was properly introduced to the space shuttle. Of course the same micro-gravity issue that had occurred with the plastic bags still remained: nothing would fall "down", so the shuttle designers developed an ingenious system of putting small jets of air from around the inside of the bowl to force the solid waste down into a blender that would break it down into smaller chunks for a waste processing unit to sanitise and store. If you ever hear someone questioning how brave astronauts *really* are, just remind them they have to dangle their sensitive bits over a blender when they crap.

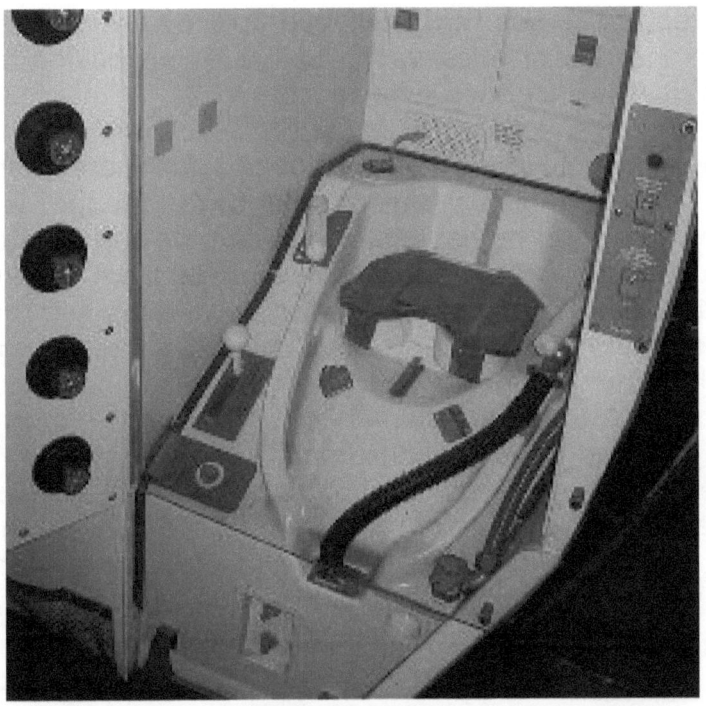

I mean, this looks simple to use...right?
[Credit: NASA]

The air jets and blender were anything but perfect. Astronauts from the shuttle program still regularly reported the occasional "floater" escaping the bowl that had to be chased back into the shiny steel deathtrap from whence it came. Similar systems continue to be used on the International Space Station and are connected to the onboard recycling system to extract moisture; but while things have radically improved since the space shuttle, there is still a real potential for it to get messy if you don't aim properly. Before launch, astronauts and cosmonauts have to effectively be "toilet trained" again, this time using a specially-built space toilet replica with a camera fitted inside the bowl. The training toilet isn't actually "used" as such, but it gives future residents of the space station a chance to practice positioning themselves using a nearby big screen tv to see how well lined up they are, so they can reduce the risk of missing the bullseye and having to deal with a floater once they're in space.

On the way to Mars you'll be doing a lot more than just extracting the moisture out of your turds, though. It turns out that the food humans eat *and* what we turn it into with our bodies are *both* great at shielding us from the kind of radiation astronauts are exposed to while travelling through interplanetary space. With that in mind, it's been suggested that the inner hull of the spacecraft could be lined with food to supplement the radiation shielding during the journey to Mars. As that food is consumed, the solid waste that's created is freeze-dried and vacuum-packed, then placed back against the inside of the hull to fill the gap. So there is every possibility that I will arrive at Mars in a spaceship lined with shit. I'm not saying I wouldn't have still signed up to go, but I feel like there probably should have been some sort of disclaimer when I was.

Back on the liquid waste side of things, nothing has really changed much for men from the old "hose and space rainbow" of the Gemini and Apollo days. However, given that women were being trained to fly on the space shuttle too, and that the NASA engineers designing it were almost exclusively men with a clear drive to solve human problems with entirely horrifying mechanical solutions, the first female astronauts for the space shuttle had casts of their vulvas taken so custom-made fittings could be connected to the urine hose system. These were never *actually* used since women are perfectly capable of peeing in a funnel as well as men can, but apparently some of the astronauts who had these casts done still keep them as quaint little paperweights. NASA engineers excelled again in their knowledge of the female body just before Sally Ride became the first American woman in space in 1983. Concerned that her period might start during the mission, they asked Ride if 100 tampons would be enough for the week-long mission. She assured them it would be.

All of this complex space toilet mischief becomes a lot simpler once we're actually *on* Mars though. With 38% of the Earth's gravity, things *will* fall and break off much the same way they do on Earth except they're going to do it nearly one-third slower. So while the first toilets on Mars won't need the same terrifying combination of vacuums, jets, and blenders required in microgravity, they will now have a slow-motion function you never asked for. There's also plenty of dirt to cover the living habitats with to provide radiation protection, so instead of radiation protection, we'll probably wind up using waste as "humanure" to grow crops the same way Mark Watney did in The Martian. Wonderful.

Rub-a-dub-dub
The other question raised in all of these horrifying toilet adventures is how do you keep yourself clean? If you make a mess in the space toilet you're going to want to scrub away the despair afterwards, but how do you have a shower in space, let alone cry in one? Water isn't going to jet out of the showerhead for long before it starts to form a massive ball in the middle of your spaceship, which I think we can all agree isn't great. The US space station "Skylab" had an on-board shower, although I use the term "shower" rather loosely: it was more like a big bag with a flexible rubber tarp over the top of it that astronauts could stick their head or body out of without releasing water and foam throughout the space station. There was enough water for each of the three crewmen on board to have a shower once a week, but the whole debacle would take about two and a half hours for each of them.

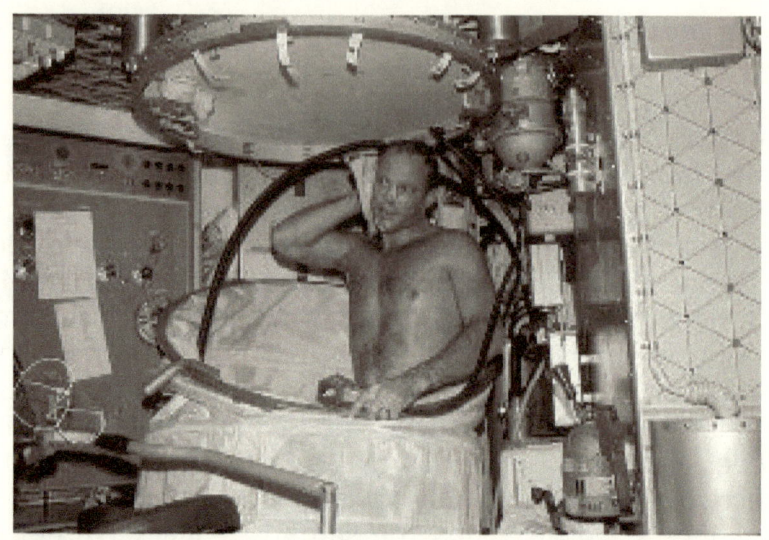

Jack Lousma sharing a candid shot for the ladies during Skylab 3
[Credit: NASA]

There's two reasons why humanity basically stopped even *trying* to shower in space: firstly, you really don't need it. Your clothes are floating around you body rather than hanging from it like they do on Earth, so they're less likely to get trapped in some sweaty and gross part of your body. You're also generally less sweaty and gross, because moving around takes little more effort that a gentle push, which is why if you're not in the gym fifteen hours a week you start to wither away into something resembling Mr Burns. Outside the gym though, that lack of daily exertion against gravity means the amount of sweat and oil your skin produces drops significantly as well, so your skin isn't producing as much stuff that could make your free-floating clothes stink in the first place. During his six-month command of the International Space Station, Chris Hadfield has said he would wear the same clothes for a week before throwing them into the waste disposal storage, and they still wouldn't stink when he did change them.

The second reason showers in space suck is the sheer cost of getting water - or *anything* - into space. While commercial operators like SpaceX are trying to radically bring down the cost of launching supplies into space, the average cost per kilogram is still about $10,000 USD. So that two-litre carton of milk you want to pour over your cereal for breakfast is going to cost a little over $20,000 to get into space, then there's another $9,000 for the 900 gram box of cereal, as well as an untold emotional cost of looking like an absolute moron trying to pour milk in microgravity (make sure you livestream it for me). With a ten-minute shower on Earth requiring approximately 20 litres of water - water that could be drunk, used to rehydrate food, or even used to generate oxygen - it's far too much of a heavy and precious commodity in space to be using it to just clean yourself. These days on the International Space Station you just use a small damp towel when you need a bit of a cleanup. It's not glamourous, but then neither is crapping into a blender.

While the cost of getting things to Mars is considerably higher than getting them to low Earth orbit, at least we'll be able to have showers again when we get there. Rather than taking water with us all the way from Earth to Mars, we'll extract it from the moisture-rich soil to produce water for drinking *and* showering. Mars' gravity also means that while water will form larger drops that won't fall as quickly as they would on Earth, they *will* still fall the right way. So a shower might be weird, but it would work. Water from the Martian soil will also be used to produce the oxygen in your shiny new Mars crew habitat too, and that's where things switch from being hygienically inconvenient into life-threatening.

To Oxygen, or Not To Oxygen, That Is The Question

Once you step outside the relative safety of a spaceship, most people assume it's the whole "not having any oxygen" thing that makes space dangerous. The reality is most people can go several minutes without oxygen before permanent brain damage starts, and another 5-10 minutes before they actually die. Air is about 21% oxygen and we need at least 16% oxygen to remain conscious, but the higher up you go, the thinner the air becomes. So while you might be breathing 21% oxygen on the top of Mt. Everest, you're only getting the equivalent of 10.5% Oxygen. Where you'll probably pass out and die. Don't worry about panicking and gasping for oxygen though, because you won't. Gasping comes from excess carbon dioxide rather than the lack of oxygen, so you can suddenly pass out from a lack of oxygen without gasping for breath once. To get around this, high-altitude pilots breath pure oxygen, but even then they'll only stay conscious below 15 kilometers without a pressure suit.

Not that you really want to be breathing pure oxygen, anyway. Low-level increases of oxygen for extended periods - like breathing 50% or greater oxygen continuously for several days in a hospital bed - causes pulmonary oxygen toxicity. Often referred to as the "Lorrain Smith Effect" after the guy who discovered it, this is where the very oxygen you're breathing begins to burn your lungs. Breathe pure oxygen at high pressure - like scuba diving with pure oxygen instead of regular air - and within minutes you'll discover another of oxygen's wonders: central nervous system oxygen toxicity, also know as the "Paul Bert Effect". It's named after the French zoologist who first described the twitching, nausea and convulsions that occur when your central nervous system is being poisoned by an overload of oxygen free radicals. Excessive oxygen exposure at pressure can also make you temporarily short-sighted, but apparently no one wanted to have *that* named after them.

This is if your oxygen-filled spaceship doesn't just spontaneously burst into flames anyway. At concentrations over 40%, oxygen can cause flammable oils to spontaneously combust, while higher oxygen concentrations will make even everyday items significantly more flammable. The tragic consequences of having too much oxygen in a spacecraft were realized on January 27th 1967, when a fire ripped through the Apollo 1 capsule during a full systems test with astronauts Gus Grissom, Ed White, and Roger Chaffee trapped inside.

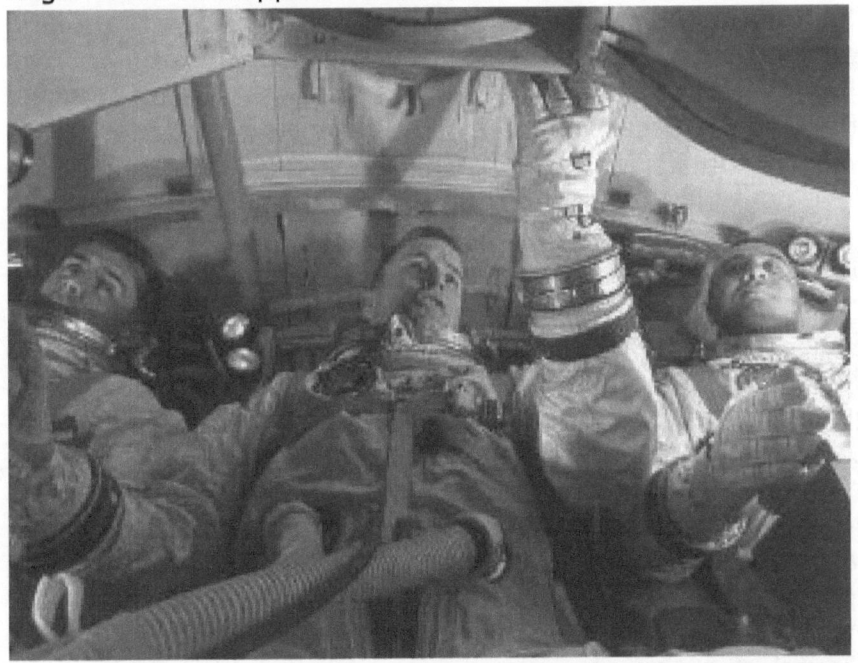

Crew of Apollo 1: (Left to Right) Roger Chaffee, Ed White and Gus Grissom
[Credit: NASA/JPL-Caltech]

With the cabin pressurised to full atmospheric pressure with pure oxygen, a spark in the faulty electrical system quickly spread fire throughout the combustible nylon interior. The intense heat rapidly increased the cabin pressure, preventing the astronauts inside from opening the inward-swinging hatch to escape while the poorly prepared support crew outside scrambled helplessly to open it from the outside. The complexity of the hatch door and the intense heat meant rescuers took more than five minutes to open the capsule, but they were far too late. There was less than 15 seconds between the first report of "fire" from the crew to their final transmission as the pressure ruptured the command module's inner wall.

The investigation of the Apollo 1 fire led to significant changes in the design of the command module, including reducing the oxygen level during launch as well as simplifying the hatch release. But the loss of Gus Grissom (the 2nd American in space), Ed White (the 1st American to walk in space), and Roger Chaffee (a decorated pilot who flew spy planes during the Cuban missile crisis) was a heavy price for NASA to pay, and a significant setback the US's lunar program.

Even if you mess up with too much or too little oxygen on the ground though, chances are that's not what will kill you quickest once you're actually in space: that honour goes to lack of air pressure. While what we breathe at sea level might feel weightless, the collective mass of gas weighing down our bodies adds up to about the same pressure as what you'd find in a football, regardless of whether it's spherical, egg, or some other stupid sportsball shape. Mostly we don't notice this weight of air unless it's moving around, for the same reason a fish wouldn't notice the weight of water: it surrounds you every moment. Drag a fish out of water though, and it'll start to suffocate. Drop the air pressure to zero on a human and they'll start to suffocate too, but what's really going to kill them is having their blood bubble while the liquid in their eyes and nose boils off...

Drop The Pressure

Most of us know water boils at 100 degrees Celsius - that's why Anders Celsius marked the boiling point "100", marked the point where water freezes with a "0", and put 99 evenly spaced graduations between them. That's also why the Celsius scale makes *sense* as a unit of measurement, rather than picking two completely random numbers like 32 and 212 just to be a jerk like Daniel Fahrenheit did. That's right: bring it on Imperial fans. You'll find no conversions for your antiquated measuring system in this book, mostly because your inability to join the 20th century is why NASA lost the Mars Climate Orbiter. Deal with it.

What you may *not* realise is that water's boiling point is directly related to the surrounding air pressure. Climb to the top of Mt Everest at 8,870m where the air pressure is about one-third of what you experience at sea level, and water boils at just 87 degrees Celsius. Another 15 kilometers up and water boils at a mere 40 degrees; and even breathing pure oxygen isn't enough to keep you conscious for long without a pressure suit. If you keep going higher, you'll reach a point at about 19 kilometers where the air pressure is low enough for water to boil at 37 degrees Celsius, which also happens to be your body's average core temperature. At this so-called "Armstrong Limit", the heat of your own body causes the fluids on your tongue, nose and eyes to literally boil off. Before you ask, no, it is *not* named after the first man on the Moon. *Frank Armstrong* was an American researcher in the 1950s who, while testing the capabilities of high-altitude pilots, pinpointed the altitude your own bodily fluids vaporise, so I'm sure he was a great guy to have at parties.

While gradually ascending into the sky until your tongue and eyes bubble momentarily before you lose consciousness would undoubtedly suck, the faster the pressure drops, the worse things get. The oxygen and nitrogen in air stay dissolved in our blood because of the ambient atmospheric pressure, so if you start *reducing* that pressure those gases are going to start coming back *out* of your blood. For oxygen it's not too bad because our body does a great job at chemically storing it in the blood as hemoglobin, but nitrogen is just there floating about ready to clog up the place. Drop the pressure faster than you can breathe that nitrogen out though and it starts to form bubbles in your blood, which you've probably guessed isn't great for the whole "blood carrying oxygen around the body" thing. It's also not great when these bubbles start to collect around your joints and cause agonising pain, or collect in your brain to cause an aneurysm and kill you. SCUBA divers will recognise this as "Decompression sickness" aka "The Bends", but you certainly don't need to be underwater to experience it: it can also occur in space. The solution in space is simple though: make sure your spaceship doesn't have any holes in it, and put your astronauts in spacesuits to make sure those troublesome gases stay in their bodies without boiling out. Although doing that can be more challenging than you might think.

(Don't) Read The Manual

In mid-1971 Soyuz 11 had just succeeded in setting two new records in human spaceflight: the longest amount of time anyone had ever spent continuously in space, as well as the very first successful docking to a space station by a space station. During their 22 days in space not only did Georgy Dobrovolsky, Vladislav Volkov, and Viktor Patsayev overcome the challenges of orbital mechanics to catch up and dock with Salyut 1, they also repaired a broken ventilation system once they opened the hatch, carried out numerous experiments, and discovered the entire station shook any time they tried to use the treadmill!

Where things went wrong was when they detached from Salyut 1 to return to Earth. Shortly after the retrorockets fired to begin re-entry the radio chatter from the crew fell silent. Mission control tried to raise them with no response, all while the capsule completed a perfect re-entry into the Earth's atmosphere - maybe it was just a problem with the radio? The parachutes deployed automatically, and the proximity-sensor controlled retrorockets fired just above the ground to soften the impact for the cosmonauts inside, but when the recovery crew knocked on the outside of the capsule there was no reply from inside. At the time the Soviet space program used a number system to report the health of the cosmonauts back to mission control: 5 being perfect health, 4 for light injuries, 3 for moderate, 2 for severe, and 1 being dead. So when the ground crew opened up the capsule and saw Soyuz 11's crew lifeless with dark blue splotches on their cheeks and blood running from their noses, they radioed through the simple message "1-1-1". Official cause of death? Hypoxia due to capsule depressurisation.

During reentry, the firing of two explosive bolts to release the descent module had jarred open a ventilation valve. The valve was between two seats and difficult to reach, so fellow cosmonaut Alexey Leonov had spoken to the crew prior to launch about closing this valve before undocking from the station just incase it failed, and advised them not to follow the official undocking procedure precisely because it didn't include closing this valve. Unfortunately the procedure *was* followed perfectly, the valve was left open during undocking, and when the retrorockets fired for reentry at 168 kilometers above the Earth the valve failed and began venting the capsule's atmosphere.

The crew probably would have been fine if they'd worn pressure suits. The problem was the Soyuz, at the time, was only designed to carry two cosmonauts in pressure suits...or three without them. In the race to put more people in space for longer than the Americans, the three crewmen of Soyuz 11 just wore simple coveralls and didn't stand a chance. To date, these three men remain the only people who have died "in space" by being above the internationally-recognised 100 kilometer altitude.

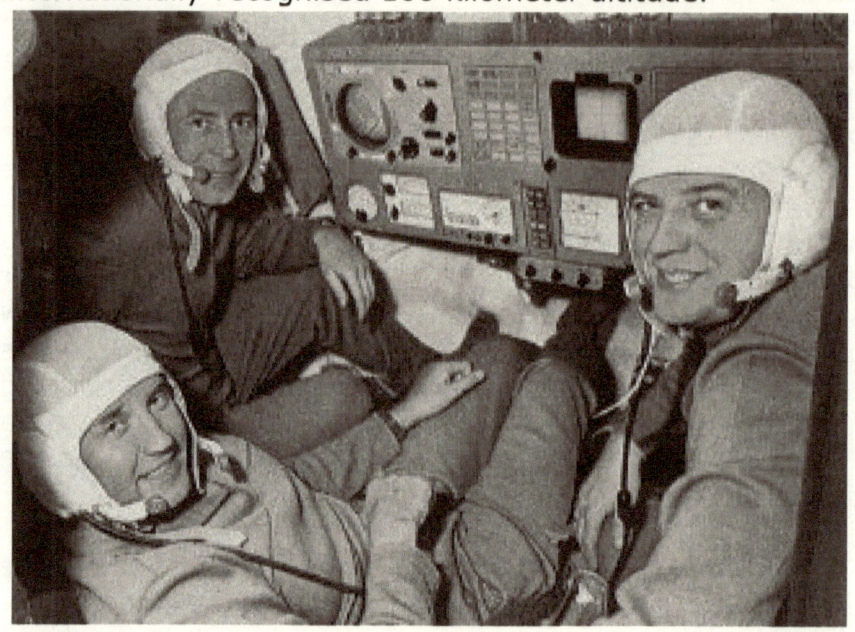

Crew of Soyuz 11: (Left to Right) Patsayev, Dobrovolsky and Volkov.
[Credit NASA]

Let's Get Spaced
So the real question becomes, "How long would you last in space without a spacesuit?". Considering all the nitrogen in your blood is going to start dissolving as soon as you hit the hard vacuum, you might think the follow-up question of, "Would it matter if you held your breath?" is a little bit irrelevant. Turns out holding your breath is actually the deciding factor between dying -fairly quickly and almost instantly.

The number one rule in SCUBA diving is "never hold your breath", and there's a really good reason that should apply to anyone about to find themselves tumbling out of an airlock into space without a pressure suit on. Anyone who's taken high school Physics was probably bored to death by Boyle's Law, where the volume of a gas is proportional to pressure and inversely proportional to temperature. On the other hand, anyone who's breathed from a SCUBA cylinder at depth then held their breath to the surface will probably have their *lungs explode* because of Boyle's Law. If you're 10 meters under the sea, there is twice as much pressure on your body as there is at the surface. At 20 meters down you're under three times as pressure, so if you take a full breath from a pressurised diving cylinder and try to swim to the surface where the pressure is at one-third of what you breathed in, those two airbags in your chest called lungs are going to try to get *three times bigger*. In diving medicine this is known as a "lung embolism" because doctors like to use fancy words like "embolism", when they really mean "a seriously bad day at the beach". SCUBA divers avoid all this awfulness by simply not holding their breath - ever. If, through some wacky hijinks, you find yourself underwater with an empty SCUBA cylinder, you'll need to perform what is known as a controlled emergency swimming ascent: swimming gently to the surface while *continuously breathing out*. Breathing out ensures your airway remains open all the way to the surface, thus avoiding lung explosion. When I've taught this to new SCUBA divers I've always told them to say "I'm a dickheaaaaaaaaaaaaad…" all the way to the surface, because 99 times out of 100 they've run out of air because they weren't keeping track of it. Submariners are also trained in the same technique, so in an emergency they can escape from a disabled submarine at depth, but if things have gone so wrong with your life you need to swim out of a submarine to survive, I can only assume you'd be saying something that sounds a lot more like "Fuuuuuuuuuuuuuu…".

In space, this problem with lung expansion is magnified again though. You might not be breathing air at high pressure like you do underwater, but when the pressure drops it goes all the way to zero. Nada. Nothing. Zip. So if you happened to find yourself about to be violently ejected from an airlock into the cold nothingness of space, unless you *want* to know what your lungs taste like, you should probably at least try to breathe out. You'll still experience about fifteen seconds of horror as your tongue, eyes, and blood start boiling before you pass out, but if you've breathed out at least your lungs won't instantly explode. Once you pass out, though, you're saved the trauma of watching your entire body swell to about twice its normal size as the gases in your blood and muscles try to bubble out. After about 30 seconds in the inky abyss your heart will stop, and then there's probably less than a minute for someone to get you back under normal pressure before it's game over for good.

In case you're wondering, this is only marginally less severe on the surface of Mars. While there *is* an atmosphere, at just 6 millibar it's about 1% of what we have at sea level on Earth, or the equivalent of being at an altitude of 30 kilometers. That's about 2.5 times higher than a commercial airliner and well above the point your bodily fluids boil at 19 kilometers up. So if you're up for a stroll on the Martian surface, make sure you're snugly sealed up in a Mars suit when you do it.

Spacewalking Bear-Fighter
When Alexey Leonov gave the Soyuz 11 crew advice on changing the undocking procedure, he'd already had his own close call with depressurisation in space. Six years earlier Leonov had become the first person to ever walk in space, and after having a bit of an issue getting back inside the capsule, very nearly became the first person to die in

space at the same time. He also probably fought a bear in the same 24 hours, but we'll get to that.

In March 1965 Alexey Leonov and mission commander Pavel Belyayev were launched into space, aboard Voskhod 2 from Baikonur Cosmodome; the very same place Russia still uses for space launches today. Everything was going perfectly as they completed their first orbit, deployed the inflatable airlock, and Leonov stepped into it. Of course, the moment he stepped *outside* to begin the spacewalk things immediately turned into a life and death comedy of errors. The capsule camera Leonov deployed only captured a few images before it jammed, while Leonov quickly started to cook and his suit ballooned in the heat of the sun. Leonov spent nine minutes trying to enjoy the view - sweating out five litres of fluids and getting heatstroke - before finally heading back to the airlock. He hadn't been able to take any photos with his other camera because the suit inflated so much he couldn't reach the shutter button, but it wasn't until he tried to climb back inside that he realised he couldn't even bend his arms or legs enough to fit in the airlock. Did Alexey Leonov - hero of the Soviet Union and general badass - let a little thing like that stop him? No. He simply *let the pressure out of his suit*. Remember that whole "pressure keeps humans alive" thing we keep coming back to? Leonov knew his solution to this minor issue would probably scare the folks at home, so he did what any emotionally unavailable man facing a serious problem does: he decided not to tell anyone.

After risking decompression sickness and a possible embolism by venting his suit to bend back into the airlock, Leonov took three minutes longer than expected to get back inside, so now there was a race against the clock. With both Leonov and Belyayev trapped in clumsy spacesuits, they now needed to seal the hatch that had jammed in the sun's heat, jettison the fabric airlock, spin the capsule around for re-entry, manually override the malfunctioning automatic landing system, and shift their equipment around to balance the capsule's centre of mass...and do it all before firing the retrorockets at a precisely-calculated time to begin their return to Earth on target. Turns out they were 46 seconds late.

46 seconds doesn't *sound* like a long time, but when you're in orbit and doing about 7.6 kilometers per *second* it tends to add up. The stabilising parachute *also* malfunctioned during re-entry, throwing Leonov and Belyayev wildly around inside just to add to the fun, which led to their capsule landing 386 kilometers off-course. So instead of touching down gently on the beautifully flat steppes of Kazakhstan, they landed in a thick Western Siberian forest. A forest full of wolves and bears during the middle of mating season: you know, *right* when they're most aggressive. To top it all off, the explosive bolts keeping the hatch in place fired as soon as they hit the ground, so the hatch they'd struggled to close earlier - the one that should keep the wolves and bears out - now wouldn't close at all.

It didn't take long for the recovery helicopters to locate the capsule, but the forest was too thick to land so they just dropped some warm clothes (clothes that then got tangled in the trees, so the crew couldn't reach them to use anyway) and flew off again. The outside temperature dropped to -5 degrees Celsius and the capsule heater wouldn't work, so Leonov and Belyayev stripped naked to wring the sweat out of their underwear so it wouldn't freeze, and then put it back on. An even bigger issue than freezing was the bears and wolves that had picked up their scent and were now prowling around the capsule. Leonov explained in later interviews that they were never worried because they, "had a pistol with plenty of ammunition". Apparently becoming the first person to walk in space wasn't enough for Alexey Leonov, he had to do it while nearly dying of heatstroke and hypothermia before probably becoming the first person to kill a bear during a space mission.

Leonov and Belyayev were reached the following day by a recovery team on skis, but they still couldn't airlift them out. The recovery team and crew built a hut and a massive fire to stay another night, and the next morning - two days after they first landed - they all skied out several kilometers to where the recovery helicopters could finally pick them up. If you ever need a template for a badass cosmonaut, Alexey Leonov is your guy, and Pavel Belyayev is not far behind.

We are just two casual Russian spacemen, no?
Leonov (Left) & Belyayev (Right)
[Credit RIA Novosti]

Temperature

Now, you might be wondering why Leonov was doing something silly like getting heatstroke during a spacewalk, when you were sure everyone says space is ridiculously cold. Turns out in space, temperature simultaneously means *nothing* and *everything*. While the background microwave radiation may be a chilly -270 degrees Celsius, the lack of an *atmosphere* to carry that heat away causes huge issues for spacecraft, satellites, and badass Russian dudes floating outside a spaceship 300km above the Middle East.

The wiring and transistors in modern electronics provide a tiny amount of resistance to the electrical power running through them, and this gets turned into waste heat. Computer processors work more effectively when they're kept cool though, and here on Earth the cheapest and most effective way to keep things cool is to use fans to circulate air. Your grandmother's computer doesn't sound like that because there's a dozen mice violently mashing away at a mechanical calculator inside: the dust filter for the case fan is probably clogged with decades of boring stories about how lazy young people are. Even without a fan to help, on Earth, heated air becomes less dense and rises the same way a hot air balloon rises above the cooler surrounding air. As it rises, cooler and denser air also rushes in to fill the area around the component, helping cool things down further. In space, though, not only is there no "up" for hot air to rise to, there's also no air to carry that heat away in the first place. Without any way to convect heat away, your only option is to conduct it into another part of the spacecraft before emitting it *very* slowly as thermal radiation. At the same time, you're also being heated by solar radiation. That same glorious sense of warmth you feel standing in the Sun on a cold winter's day can turn fatal when there's no "cold winter's day" to carry it away again and the outside of your suit heats up to 135 degrees Celsius. Alexey in his spacesuit didn't have any way to radiate that heat away, so he was essentially stuck inside the equivalent of a thermal flask, with the sun heating him from the outside while his own body heat cooked him from the inside.

Satellites and spacecraft are often covered in highly reflective materials to reduce the amount of heat they collect from the sun in the first place. That highly reflective gold foil you see on the outside of satellites is a mixture of actual gold and Kevlar: the gold reflects away solar radiation while the Kevlar adds a little protection from micro-meteorite impacts. Covering a flexible fabric spacesuit in gold isn't particularly practical though, so their outer shell is made white to reflect as much sunlight away as possible.

Spacecraft and modern spacesuits are designed to conduct the heat generated by the people and electronics inside away, to large metal fins or pipes filled with ammonia, releasing that heat energy as thermal radiation. These systems are much slower, heavier, and less efficient than anything you might find on Earth, but they are absolutely vital to avoid everyone and everything being cooked. The International Space Station was nearly incapacitated in 2010 when an ammonia pump on one of the station's two primary cooling loops failed. With half of the cooling system? offline, the station was forced to run at minimal power with no backup for several critical systems for over a week while Tracy Caldwell Dyson and Doug Wheelock conducted three emergency spacewalks to disconnect the broken ammonia pump, fit one of the four spare pumps that were mounted to the outside of the station, and reconnect the ammonia lines to the cooling system.

The huge 353 kilogram ammonia pumps seem to be an ongoing issue for the International Space Station, with emergency spacewalks being required to repair and replace them again in 2013 and 2015. Worse still, every step of these spacewalks involves watching out for "snowflakes" of frozen ammonia leaking from the pump lines. If frozen ammonia stuck to a spacesuit and was brought back inside the station, it'd turn into a gas and poison the astronauts as soon as they took off their helmets. So any spacewalkers exposed to ammonia systems need to decontaminate their suits at the end of each spacewalk by baking them in sunlight before coming back inside. You won't hear many complaints from astronauts about waiting around outside when you consider the view though.

Tracy Caldwell Dyson looking *really* unhappy about having to wear a spacesuit
[Credit NASA]

As Alexey Leonov's case pointed out, spacesuits need some serious cooling systems to stop you sweating yourself into a puddle, but they also need to keep enough heat in so you don't freeze whenever you're in the Earth's shadow. NASA's spacesuits on the International Space Station need to be able to handle temperature changes between -160 and +135 degrees Celsius, going from the Earth's shadow into full sunlight. Even so, astronauts conducting spacewalks on the International Space Station's Canadarm2 often report having cold feet, as well as feeling vaguely ridiculous being strapped to a giant robotic arm that's hurtling around the Earth at over 7 kilometers a second.

Strangely enough, extreme temperature is less of an issue on Mars than you might expect. The whisper-thin atmosphere provides enough protection from the Sun and carries away enough heat that the surface exploration suits will only need to handle temperatures ranging between -128 and +77 degrees Celsius - a full 90 degrees less variation than those on the space station. So while Mars might have a rather brisk average surface temperature of -55 degrees Celsius, that tiny atmosphere still makes it much safer than being in open space.

Slip Into Something More Pneumatic
You can't just take any old space suit to Mars, though. Space suits have gone through generations of development since the first flight suits were made for high-altitude U2 spy plane pilots in the 1950's. In the US, the first "spacesuits" were the now-classic reflective silver suits developed for the astronauts of the Mercury program. They were refined throughout the Gemini program, before evolving into the Moon suits worn by the Apollo astronauts on the lunar surface, and eventually into the modern suits used for spacewalks on the American side of the International Space Station today. The Soviet space program went through a similar evolution, moving from simple pressure suits to protect cosmonauts from a Soyuz 11-style capsule depressurisation, learning from the mistakes made with the suit Alexey Leonov wore on his first spacewalk; to the modern Orlan spacesuits worn by spacewalking cosmonauts on the Russian side of the International Space Station. Future spacesuits worn on Mars will also have to contend with the effects of perchlorates in the Martian soil: highly toxic superoxides that will eat away at both spacesuit fittings *and* astronauts' lungs in very short order.

What all current suits have in common is that they're pneumatic, meaning they're filled with gas to keep pressure on your body, and act like a mini spacecraft that protects your squishy wet organs from the harsh vacuum of space. Pump them up with too much pressure, or overheat them, and just like Alexey Leonov's suit they become impossible to bend. Modern spacesuits operate at a pressure about the same as you'd experience on the top of Mount Everest, and use pure oxygen - with all its associated fire hazards - so spacewalkers can avoid passing out from hypoxia while they're strapped to a 400 ton space station circling the Earth every 90 minutes. Even at such low pressure, their body is still essentially trying to bend and twist a football from the inside, so while pneumatic spacewalking suits are great for keeping all the fluids and gas in your body from boiling out they're incredibly hard work to wear and completely unforgiving to the outside of your body. Astronauts regularly return from spacewalks with lost fingernails and bleeding fingers from where their gloves have worn away at the skin; and rubbing points and abrasions around the elbows and knees are just as common. That's just doing normal work in a fully functioning space suit too: it's when something in a suit breaks that things start to get really ugly. Drowning in space might sound utterly ridiculous, but in July 2013 it very nearly happened.

No Lifeguards In Space
An hour into his second ever spacewalk, Italian astronaut Luca Parmitano felt a weirdly wet sensation on the back of his neck. A week earlier, he'd experienced something similar at the end of his first spacewalk while fitting a new communications module to the International Space Station. Ground control suggested that after six hours of rattling around outside the station his helmet drinking water had probably leaked a little. Nothing to worry about! This time,

though, it felt like a *lot* more fluid, so after radioing through to ground control they told him to dump the water through a purge and continue on with the spacewalk. Minutes later, it was clear the drinking water wasn't to blame, and something much more serious was at hand.

As the fluid kept filling from the back of Luca's helmet, the weightlessness of space meant it stuck to his bald head, working its way up and over his scalp. With his spacewalk partner Chris Cassidy too far away to help and knowing there was *far* too much fluid to have just been his drinking supply, Luca activated the emergency retraction line to pull him blindly back into the station's airlock as the fluid covered his eyes and quickly started to envelop his nose and mouth. Unsure if he could open his mouth to breath, Parmitano still hoped it was only water - if it was contaminated with anything else then opening his eyes could leave him permanently blinded. As he considered venting his helmet as a final option, Parmitano clamoured blindly back inside the airlock where the crew inside repressurized and watched over a litre of water float away from Parmitano's face as they removed his helmet.

Water clinging to the inside of Luca Parmitano's helmet after his near-drowning
[Credit NASA]

All planned spacewalks were put on hold while the crew determined the problem, but when another ammonia pump failed a few months later, Rick Mastracchio and Mike Hopkins were forced to conduct a series of emergency spacewalks knowing full-well the fault that nearly drowned Parmitano hadn't been found, and they wouldn't be able to dash inside the airlock if it occurred again because of the ammonia exposure. The helmets were fitted with emergency snorkels as a last resort, luckily they didn't start spontaneously filling with water, and the pump was replaced; but years later it's still unclear exactly what actually caused the flooding. So far, Luca Parmitano is the only person to nearly drown in space, but it's understandable that a lot of people preparing to go to space want to wear something a little better than a 145 kilogram spacesuit designed in the 1980's that you might also drown in.

Crushingly Sexy Suits
To get away from the issues with bulky, heavy, and stiff pneumatic suits, NASA has been looking at "mechanical counter pressure" suits since the early 1950's. Rather than maintain a body of gas around the wearer that provides the necessary pressure to keep their body from swelling up, these "skin suits" provide the necessary pressure to our bodies by squeezing directly in on them. While a g-suit operates in a similar way, by compressing a fighter pilot's legs during a hard turn to stop all the blood draining away from their brain, a mechanical pressure suit used above 15,000 meters would need to squeeze their entire body. That's going to be an issue for your face, and unless you're into that kind of thing, I doubt your crotch will enjoy it for long either.

That's the problem with mechanical counter pressure suits: they still have to squeeze you *everywhere* with that pressure. While MIT's modern "Biosuit" design has gone a long way towards comfortably providing these kinds of pressures, they still have to be literally skin-tight and custom-designed for each wearer. Not only does this make it an absolute nightmare to take on and off, but you *really* also don't want any part of the suit to fold or pinch. There's also the issue of being isolated on a distant Martian outpost and your figure-hugging spacesuit being simply too sexy for the rest of the crew. Although I'm sure you could dial things down for your fellow Martians by wearing a mouldy old tracksuit over the top while eating everyone's ice cream ration through a straw and talking constantly on the radio about your ex.

Radiation
After running out of oxygen, radiation is usually the next thing everyone freaks out about. Humans tend to fear things they don't understand, especially if they can't see it or it might harm them, so mysterious invisible rays that can kill you are practically the boogeyman for our lizard brains. Also, I looked up "bogeyman" to find something crosscultural, and it turns out no matter where you're from adults love to terrify kids with stories about monsters in the dark - we are so screwed as a species.

On Earth we're exposed to relatively small amounts of radiation every day: the microscopic amount of radiation you're exposed to from sunlight, the tiny amount of potassium-40 in a banana, or the fraction more when flying long-distance or getting a chest X-ray. The mix of oxygen and nitrogen in our atmosphere blocks most of the radiation from the sun, but lets through certain frequencies. Visible light is one band of frequencies that gets let through, which is why we've evolved to see with it, but even that is affected by the atmosphere. The sky is blue because higher-frequency blue light is scattered more than lower-frequency red light. It's also why at sea level the vast majority of ultraviolet light is scattered by the atmosphere, but can still give you a sunburn. At higher altitudes there's less atmosphere to protect from the Sun's radiation, which is why mountain climbers and pilots are particularly susceptible. Those big aviator sunglasses aren't purely for show.

This is all relatively insignificant compared to what is going on once you leave the comforting embrace of our atmosphere and head out into space, where you're bombarded by the full spectrum and strength of the Sun's radiation: X-rays, Gamma radiation, even microwaves. Because the Sun is essentially a 1.4 million kilometer-wide, unshielded, nuclear fusion reactor, it's also throwing out a soupy mix of high-energy helium, electrons, and beta particles along with all that electromagnetic radiation as well. The Earth's rotating magnetic field has created the two "Van Allen" belts that shield us from the harshest elements by trapping the high energy particles and deflecting the rest towards the poles. These deflected particles then slam into oxygen and nitrogen in the upper atmosphere and glow in the night sky to create aurora. You'll have to travel to the high Arctic at the right time of year to see the Aurora Borealis, or the Antarctic to see the Aurora Australis, but you can see either of these stunningly beautiful phenomena fairly regularly if you're on board the space station.

Of course this picture completely sucks in black and white
[Credit: NASA]

Even if you were lying naked on the back of an elk in the Finnish Lapland under the aurora borealis, the solar wind isn't going to even warm you up, let alone kill you. Hypothermia probably *will* kill you though, so you might need to turn that elk into a Tauntaun and climb inside it until morning. Also, please stop riding elk around Finland while you're naked - it's weird, and you're scaring everyone who came to see Santa.

That shimmering glow of the aurora can cause absolute havoc for radio communications, and in space the solar radiation that creates it can outright kill satellites. A peak in solar wind overloaded the sensitive electronics on board the Galaxy-15 telecommunications satellite in 2010 and sent it rogue. While it was still transmitting data, this "zombiesat" wouldn't respond to commands to correct its orbit, started to threaten other satellites as it drifted, and generally turned from upstanding orbital citizen into a menace for the entire geostationary neighbourhood. Eight months after they lost contact, ground controllers finally managed to re-establish communications and reboot its systems, but not before the incident had cost the manufacturer more than $2.5 million US as Galaxy-15 was threatening six other satellites and interrupting maritime weather forecasting.

Just after Apollo 16 returned from the Moon, a massive solar storm erupted, generating the highest class of solar flare and throwing billions of tonnes of charged particles at the Earth in what's known as a Coronal Mass Ejection. Had Apollo 16 launched later, or Apollo 17 earlier, either could have been caught in this huge burst of electromagnetic radiation and charged particles, with the crew receiving a significant but non-fatal dose of radiation. About every eleven years the Sun goes into a cycle of intense activity, consistently spitting out huge x-ray flares and coronal mass ejections as the intense magnetic field lines on the surface violently twist and reconnect before quieting down again.

All of this is fairly standard solar weather, but about once every 150 years the Sun decides to put on a real show. On September 1st of 1856 an amateur astronomer by the name of Richard Carrington was conducting solar observations from his observatory in Surrey when he witnessed two particularly bright beads of light over the sunspots he'd been recording. Within a minute they'd weakened, but Carrington knew he'd just witnessed an unexplained yet extraordinarily powerful solar event. Less than 12 hours later, a tidal wave of charged particles thrown by these cataclysmic solar explosions hit the Earth's atmosphere, causing electrical generators to burn down, telegraph lines to spark and electrocute telegram operators across the US, and creating auroras across the entire planet.

If a so-called "Carrington Event" occurred today, it would be a global disaster. It would almost certainly wipe out satellite communications and navigation systems globally, impacting everything from banking transactions to passenger jets. Unshielded electronics - such as your mobile phone and laptop - would likely be destroyed by the initial burst of energy, and the power grids would be hit particularly badly. An assessment by Lloyd's of London in 2013 estimated that if the Carrington Event occurred today, it could cause up to $2.6 *trillion* US dollars worth of damage to the electrical grid and carry a potential death toll in the tens of millions in North America alone. For perspective, if regular solar radiation is like an irritating four year-old throwing marshmallows at you, then the coronal mass ejections during the 11-year solar maximum is like Jesus from The Big Lebowski hurling a bowling ball at you: it's going to hurt, but chances are it won't actually kill you. Comparatively, another Carrington Event would be like *Godzilla throwing buildings at people*. We're not talking about the 1998 Godzilla with Matthew Broderick being an emotional simpleton, either. Nor do I mean the 2014 remake with Godzilla trying to cock-block those two nuclear flying things. This is going to be some hardcore, old school, 1950's "man in a rubber dinosaur suit stomping and burning down a tiny Tokyo" *gojira* turning up to cancel Christmas.

Bow before your lizard god, spikey one!
[Credit: Public Domain "Godzilla Raids Again"]

Another Carrington event *did* actually happen in 2012, but thankfully it was the *other* side of the Sun that exploded so Earth wasn't in the firing line. Solar events on this scale happen *on average* once every 150 years, but there's still a 12% chance of another one hitting Earth in the next decade. I don't want to get all tinfoil hat doomsday-prepper on you, but it's probably easier to tell your kids the Purge movies are a documentary series and save them some confusion later.

Shields Up
Obviously our intrepid Martian colonists are heading *away* from the Sun and its temper tantrums, but even without a huge solar event the crew will still receive a 386 millisievert dose of radiation during a 210 day trip to Mars, about a third of the lifetime limit according to most space agencies. If there's a huge coronal mass ejection heading towards them though, they'll want all the extra shielding they can

get. The basic idea behind protecting people from radiation is to put as much distance and radiation-absorbing *stuff* as you can fit between the people and whatever is emitting the radiation. On Earth, that usually means putting sheets of lead around the nuclear reactor you've built miles from anything else, or putting lead-crystal glass between you and the X-ray machine that's at the far end of the hospital. Not only is a spaceship made out of lead going to be too heavy to get off the ground, but the type of radiation we're dealing with in space means lead shielding would actually make the crew's radiation exposure even *worse* than if they had no shielding at all. While the continuous solar wind adds to an astronaut's radiation dose, and solar flares and coronal mass ejections add a high-level but short-term risk, the biggest source of radiation for future Mars colonists are particles from beyond our solar system. Galactic cosmic rays are incredibly high-energy particles travelling near the speed of light that astronomers suspect come from collapsing stars, supernovas, pulsars, and all sorts of other weird interstellar sources. While our atmosphere is thick enough to scatter most cosmic rays, Apollo astronauts on the way to the Moon saw them every few minutes as bright flashes of light in their eyes while they tried to sleep.

Cosmic rays strike with such force that when they hit something, they'll create a shower of smaller particles. When that "something" they hit is especially dense like lead, it creates a shower of even *more* dangerous particles on the other side. Even aluminium - the mainstay of modern spacecraft - can create a multitude of highly damaging particles inside the crew capsule that are far worse than just being hit by the cosmic ray itself. With cosmic rays it's better to have *no* shielding than it is to get it wrong. It turns out the best thing for safely scattering cosmic rays is the lightest of all the elements: hydrogen. While there is plenty of research going in to using plastics with a high hydrogen content as spacecraft radiation shielding, there's something else on board that we'll need on the way to Mars that has *plenty* of hydrogen in it: water. Storing our drinking and waste water in compartments around the hull helps scatter both solar radiation and cosmic rays and reduces the crew's radiation dose during the journey. If there was a coronal mass ejection on the way, the crew would take shelter inside a heavily-water shielded chamber in the centre of the spacecraft, with four people squeezing into a space the size of a toilet cubicle for up to two weeks while waiting for it to pass. Even if you're not worried about sharing a spaceship the size of a school bus with three other people for seven months, I doubt two weeks of crapping into one of those plastic bags while you're centimeters away from the rest of the crew in a radiation chamber would exactly be a career highlight.

In fact, wouldn't it be a whole lot easier if you just *slept* inside that radiation chamber the whole way to Mars? We've seen hibernation pods for long-duration space trips in films like Alien and Interstellar, and reality is fast catching up with Hollywood. Clinical animal trials at John Hopkins University suggest that it might just be possible to keep humans alive in a form of hibernation for up to 30 days at a time. Trials with monkeys and mice have shown that the body will put itself into a protective sleep mode if you cool the core temperature down from the normal temperature of 37 degrees Celsius to a chilly 20 degrees, and you could potentially stay asleep like that for a full month. So maybe you set course for Mars, everyone has a month-long nap, stretches their legs for a day or so, and then beds down for another month. Or maybe two of the crew sleep for a month while the others keep an eye on things, and then they swap over for the next month. Either way, having at least some of the crew in hibernation means they'll use less oxygen, food, water, and energy while they're sleeping in the part of the spacecraft with the best radiation shielding. We're not there yet, but sleeping your way to Mars definitely looks promising.

Radiation gets a little easier to deal with once you've landed on Mars too. Curiosity's "Radiation Assessment Detector" (aka RAD) measures approximately 30 microsieverts an hour, or about three-quarters of the dose you get while flying from Los Angeles to New York. All that nifty Martian soil you're surrounded by is great for blocking cosmic radiation, so if you cover your habitat with about five meters of it, the radiation exposure inside drops to the same level we get every day on Earth. You won't want to stay inside all the time, but even if you spent on average of an hour a day outside the habitat it'd still take about 60 years to reach ESA's career radiation limit. If you're *really* worried about the radiation, you could also just stay inside during the day and only come out at night. Like a vampire, living in your underground crypt. On Mars.

Zimmer Frames In Space
If we're exposing people to cosmic radiation and then shoving them underground on another planet, you have to wonder how long are we actually expecting them to survive. Turns out that as long as they don't get smacked by a once-in-150-years Carrington-type event on the way there, our intrepid Mars colonists will probably live longer than they would if they stayed on Earth! Sure they're being sent into conditions no humans have ever experienced and exposed to radiation during the trip, but that radiation is only a third of their lifetime allowable limit. Once they're on the surface of Mars they'll be breathing purified and carefully monitored air inside the habitat, eating an incredibly clean and mostly vegan diet, drinking highly purified water, and regularly having health checks to research the impact that living on Mars is having to their body.

They'll also be living the rest of their lives in a third of Earth's gravity, so their heart and joints don't have to work as hard, which is going to be great if we turn Mars into a retirement home. One idea that is regularly thrown around is to send older people to Mars instead of the young upstarts you might expect. While the zany rules of reality TV usually dictate that everyone on screen needs to be young, and fighter pilots are often flying a desk by the age of 35, space exploration values the experience brought by a slightly older vintage. The average age of a US astronaut isn't 25, it's 45. Neil Armstrong was one of the youngest members of the NASA astronaut corp when he set foot on the Moon in 1969 at the age of 38. The oldest spacewalker to date is Pavel Vinogradov who, at 59, conducted his seventh spacewalk to install an experiment on board the International Space Station. Peggy Whitson first flew to the International Space Station when she was 42, and commanded it for the second time when she was 57. The patience, caution, and maturity of older astronauts doesn't dampen their spirit of exploration, only gives them then added bonus of having the experience to recognise risks and dangers before diving headlong into a potentially fatal accident through haste. There is also a biological benefit to sending folks with a few extra years behind them too, because they have a lower risk of developing cancer from radiation exposure. With their cells slowing down and dividing less frequently, any potential Mars colonists that has not had cancer before the age of 55 is a third less likely to develop a tumor when exposed to the same level of radiation as those younger than them. That 38% gravity also means you can keep working elderly Mars colonists *well* past retirement age because the loading on their bones and joints is nearly three times lower than it would be on Earth.

Of course, just because you *can* send older people to Mars doesn't necessarily mean you'll want to. You still have to balance the benefits of experience and reduced cancer risk against the risk of illnesses like Alzheimer's and dementia. Perchlorates in the Martian soil might also make older colonists more susceptible to lung-diseases such as emphysema, with less capability to recover than a younger crew would have. There's also always the concern that as people age they'll start to tell really boring stories, smell faintly of parmesan cheese, and rant about how things were cheaper when they were young and living on Earth.

Just like an old person on Earth, getting to Mars probably means you've peed and vomited out most of the fluids in your body, had your spine stretched while going blind, probably been chased around your spaceship by some mutant space virus been threatened with drowning, blood bubbling, or just bursting into flames on a spacewalk; and lined your spaceship with bags of shit for seven months…all so you can live underground with three other mole people for the rest of your life on a planet that is trying to kill you. After all that, you have to ask yourself the question: *why in the name of Great Googly Moogly would you want to go to Mars?*

Martian Mind

On June 2nd 1957 Joe Kittinger was on top of the world. Literally. Hanging from a high-altitude balloon nearly 30 kilometers above the Earth, Kittinger was on board the first of the US Air Force's three "Man High" balloon flights designed to test high altitude equipment and see what cosmic rays would do to a human. You'd assume that the thought of being exposed to radiation like that might be offset by the extraordinary view that you'd get at that kind of altitude - the glow of the Earth and its atmosphere below you with only black sky above - but it wasn't. Kittinger was jammed inside a tiny, claustrophobic aluminium capsule for over twelve hours without a window, his only connection to the outside world an onboard radio. A radio that broke about 20 minutes after the capsule launched.

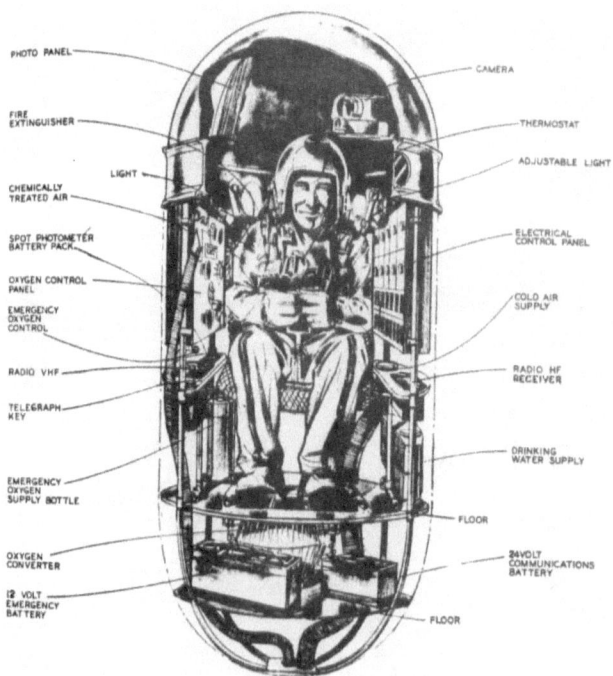

Does this look pleasant to you? The guy is *kind* of smiling, right?
[Credit: USAF]

Kittinger could still receive voice transmissions but could only transmit back using simple Morse code, making communications painfully slow. Determined not to abort, Kittinger proceeded up to 30 kilometers above the Earth and started taking barometric measurements and performing experiments to prove that the balloon system worked...when it became very clear that it didn't. In less than two hours, the capsule had somehow lost 60% of its liquid oxygen supply and it was still dropping fast, so ground control ordered Kittinger to abort and start the descent back to Earth. From 30 kilometers up, drifting through the middle Stratosphere inside a tiny aluminium tube Kittinger slowly tapped out his reply in Morse code: "COME-UP-AND-GET-ME".

Ever since the dawn of the space age, psychologists have been concerned about "Breakaway Syndrome", the theorised risk that astronauts on a mission will psychologically disconnect from Earth and try to leave it all behind for the quiet darkness of space. You've probably experienced the same sensation when you've been listening to the tax office's hold music for two hours and an automated recording says, "Your call is important to us" for the 38th time. Joe Kittinger wasn't experiencing Breakaway Syndrome, though. He was just being a smartass. Kittinger would go on to test parachute systems for pilots flying high altitude aircraft, like the X-15, and set numerous parachuting records by jumping out of an open-air gondola over 31 kilometers above the Earth in a pressure suit as part of Project Excelsior for the US Air Force. But things would continue to go dangerously wrong. During a warm-up jump from 23 kilometers, Kittinger had parachute lines wrap around his neck, choking him as he was thrown into a 120-rpm flat-spin until the reserve parachute automatically deployed several minutes later, saving his life. Even on the record-setting Excelsior III jump from 31 kilometers up,

Kittinger had one of his pressure gloves fail during the ascent, which caused his hand to painfully swell to twice its normal size. Kittinger said nothing to ground control about it until he was right on the edge of the gondola, ready to jump.

It takes a special kind of crazy to jump out of a balloon with just a parachute when you're up high enough to have your blood boil if the rest of your pressure suit fails...especially when you did the same thing a month earlier and the parachute tried to choke you to death on the way down. But those selected for the early space programs were pretty unique, even if they seemed nearly identical.

The Stepford Astronauts

When NASA first selected the original "Mercury 7" astronauts in 1959, they all came from a single branch of the military: test pilots. At the time, President Eisenhower, being a former 5-star General during World War 2, viewed spaceflight as an extension of the military's experimental aircraft test program so he insisted that only active test pilots from the Navy, Air Force and Marines could be selected as America's first astronauts. The public argument was that with their quick thinking and ability to remain calm under pressure, test pilots were already psychologically prepared to handle the extraordinary hazards of early space exploration. The thought of a few brave military men venturing into the unknown immediately set an attitude across the US that only the very best *white male* test pilots could be astronauts. No women, no people of colour, no diversity: just seven straight white guys with "perfect" military records, all aged between 25 and 40 years, all with a Science or Engineering degree, all less than 1.8 meters tall, and all from one very narrow and hyper-competitive branch of military aviation. Even Neil Armstrong was initially rejected because he'd left active military duty in 1953 to get his Aeronautics degree and work for NASA as a *civilian* test pilot, while Buzz Aldrin was ineligible because he'd chosen to write his doctorate on how to dock spacecraft in space instead of pursuing test pilot opportunities.

Joe Kittinger was one of the few people who *could* apply to the Mercury program but decided not to as he was having too much fun jumping into the sky and breaking records. He probably also didn't want all the hassle of having to constantly maintain a brave, faultless, All-American hero facade either. While the US was showcasing their smiling, crew-cut, and perfectly polished astronauts, the Soviet Union was taking a very different approach to cosmonaut selection. While cosmonauts were still generally selected from the Air Force, and the state-controlled media still loved to glorify their successes, the Soviets selected their earliest cosmonauts more for their personal background and support for communism. Yuri Gagarin was a relatively inexperienced pilot in comparison to many other candidates for Soyuz 1, but he'd been a country farm hand before learning to fly. As an inspiring example of what humble everyday Russians are capable of, Gagarin's journey from farm hand to first man in space was far more important than his flight experience. In Valentina Tereshkova's case, she had *no* flight training: the first woman in space was an amateur parachutist who worked in a cotton mill and had to be temporarily inducted into the Soviet Air Force to become a cosmonaut. For the first decade of human spaceflight, both the US and the Soviets could get away with selecting crews based on their military flying credentials and the political story they helped create, mostly because the crews really weren't in space for long each mission. Apollo 17 was the longest mission of the US's lunar landing program, but even that was less than two weeks. Obnoxious, arrogant, and hyper-competitive personalities common among fighter pilots could be overlooked because the only real requirement from the crews on a short mission was that they could follow instructions and act quickly in an emergency. There simply isn't time to do much besides what mission control is telling you to do at every moment of the day. But even the most highly trained and dedicated people can only put up with being constantly hounded for so long, and that hardline military approach began to unravel as soon as the US and the Soviets started sending people into orbit for months at a time rather than just days or weeks.

Orbital Mutiny
With Saturn V rockets left over from the cancelled Apollo 18, 19, and 20 lunar missions, NASA converted one into an orbital workshop and used another to launch it into low-Earth orbit. Skylab had huge issues when it first launched: a failed heat shield raising the internal temperature dangerously high, toxic gases filling the living habitat, solar panels jamming, and issues with the first crew even being able to dock to the station. The first three-person crew managed to fix the biggest issues during their 28-day mission, while the second crew spent 60 days on board documenting the changes microgravity was having on their bodies. It wasn't until the final Skylab crew started their *84 day* mission that things went really pear-shaped. You might not recognise the names Gerald Carr, Ed Gibson, or Bill Pogue because NASA doesn't like to talk about them much anymore, but they're the guys who staged the first fully-fledged mutiny in space.

Final Skylab Crew: (left to right) Gerald Carr, Ed Gibson, and Bill Pogue

Mission control had been hammering the all-rookie crew since their arrival, pushing them well beyond the already brutal 16-hour work days while they were still trying to recover from space sickness, something ground control discovered by secretly eavesdropping on the crew with onboard microphones. While ground control didn't think they were working hard enough and regularly ordered them to work through their meal times to make up for it, the crew complained openly that they were being overworked. Over six weeks the tension built until a few days after Christmas, crew commander Gerald Carr sent mission control an ultimatum:

"We need more time to rest. We need a schedule that is not so packed. We don't want to exercise after a meal. We need to get things under control."

Ground control didn't seem to care and told the crew they needed to meet the schedule. So the next morning Carr told ground control they were turning off the radio for the day and that the folks on the ground would just have to deal with it. The Skylab crew had a wonderful 24 hours staring out the window and catching up on everything they'd fallen behind on while mission control impotently fumed. The next morning the crew switched the radio back on and everyone had a nice discussion about how the workload would be reduced for the rest of the mission and the crew would now be given a list of things to achieve each day instead of having every second scheduled for them.

This lack of trust between the crew and mission control has been an ongoing issue since the start of human spaceflight. During the US's first orbital flight, mission control suspected the heat shield on Friendship 7 had been dislodged, but rather than actually *tell* John Glenn he might burn up on reentry they just asked him not to jettison a parachute pack that they hoped would hold the heat shield on if it *had* been dislodged Turned out it was just a faulty sensor. Even during the Apollo program, mission control still didn't trust their astronauts. Apollo 10 was a complete dress rehearsal for every aspect of the first lunar landing, minus the actual landing. Concerned that Tom Stafford and Gene Cernan might try to land anyway, NASA's engineers didn't completely fill the fuel tanks so if they did land, there wouldn't be enough fuel to leave the Moon again. In fairness Cernan *was* a bit of a cowboy, and even though he got to be the last person to walk on the Moon three years later for Apollo 17, he probably *would* have landed during Apollo 10 if they'd given him enough fuel to do it. To go all the way to the Moon, float just 14 kilometers above the surface, then have to come all the way back again? Even knowing there wasn't enough fuel, it still must have been awfully tempting.

Flight surgeons were often a particularly disliked part of the ground staff during the US's lunar program, because in the eyes of these hyper-competitive astronauts it was the flight surgeons who were constantly looking for some physical fault in their bodies that would stop them getting to the Moon. It didn't help that the chief of the astronaut office was a medically-grounded and unflown astronaut either. Deke Slayton had passed all the gruelling medical examinations to be selected as one of the original group of seven Mercury astronauts in 1959, and was set to become the second American to orbit the Earth. But in 1961, surgeons detected a slight abnormality in his heart rhythm and he was grounded by NASA and the Air Force indefinitely. Without his own opportunity to fly, Deke became the champion of the astronaut corp, managing much of the crew training and selection for over a decade during the US's Gemini and Apollo programs, fighting battles on the ground with mission control on behalf of the guys in orbit, and playing a deciding role in selecting Neil Armstrong as the first person on the Moon. After the lunar program ended, Deke was finally allowed to go to space as part of the Apollo-Soyuz Test Project in 1975: five years after his heart murmur had disappeared and *sixteen years* after he was first selected as an astronaut. At 51 years of age, Deke Slayton set a new record for being the oldest person to fly in space, but the flight surgeons still made him take heart medication with him. Just in case.

Shifting To Shuttle
The US's attitude of only sending steely-eyed missile men into space didn't even begin to change until the advent of the space shuttle. Promoted as a bus to space that would make trips to orbit routine, the shuttle program meant that after 20 years of its astronauts being a military "boys club"

NASA suddenly needed to open up to sending scientists, doctors, and civilian contractors into space. Worse still, some of those now going to space might even be - gasp - women! They still wouldn't be spending more than two weeks in space at a time, but rather than launching three nearly identical military dudes to the Moon, NASA was now planning to regularly jam *seven* highly accomplished individuals from diverse backgrounds into a cramped micro-gravity environment...and tell them to work together without killing each other or destroying billions of dollars of equipment. At least the shuttle had a toilet that mostly worked though. Mostly.

With all of this diversity flooding into their astronaut corp, NASA really needed to take a closer look at the psychology of the people they wanted to send into space to make sure they could play well with others. Astronaut candidates would have their mental health and thought process interrogated through hours of interviews, while serving shuttle astronauts remained under the suspicious eyes of psychologists to see how they communicated and handled stress; enough to actually send you crazy if you weren't already. That attitude of 'only the best and brightest go to space' still prevailed, so while many shuttle astronauts were still coming from test pilot and military aviation backgrounds, now the world's leading scientists, engineers, and medical researchers were cooped up together with them. The space shuttle's cabin may have had over 74 cubic meters to move around in, but I'm sure it would have felt pretty cramped with some of those egos.

Compared to just two weeks on the space shuttle, six months on the International Space Station is a whole other bag of burritos. It might be called the "International" Space Station, but it's essentially still spilt down old Cold War lines with only a claustrophobic tunnel filled with food and water linking the US and Russian halves. Separate sleeping areas, separate control systems, separate exercise equipment, even separate toilets. Those last two have caused quite a few issues through the years too, with Russian cosmonauts claiming they've been banned by the people on the ground from using toilets and gym equipment on the US side because of political squabbles over the US having to pay for seats on the Russian Soyuz to get to the space station since the space shuttle was retired in 2011. Likewise, when Russia invaded the Ukraine in 2014 and the US threatened sanctions in response, the Russian Deputy Prime Minister suggested the Americans could "bring their astronauts to the International Space Station using a trampoline". Apparently even 400 kilometers up you can't completely escape politics, but for the most part the crews on the International Space Station quietly dismiss the power plays being made on the ground. Are you really going to tell someone with a broken toilet they can't use 'yours', when there's six of you living in a sealed aluminium can circling the Earth every 90 minutes?

Even without the petty political issues on the ground , putting people into space for 6 month periods poses plenty of psychological concerns. For most astronauts the first month is full of excitement that they're finally starting something they've spent literally years training for. Similarly, the final month is usually filled with the excitement of returning to friends and family after a job well done. You'd think being able to stare out the window *at Earth from space* would be enough, but it's those four months in the middle that can leave astronauts feeling overwhelmed, depressed, and missing their friends and family. There have been numerous psychological isolation studies of crews aboard submarines, at Antarctic stations, and even specially-designed Mars simulations that have shown that many people will suffer the "Third quarter effect": you're over the halfway point, but the end still seems so far away. Different people respond to this kind of isolation in different ways, but for many it's a charming mix of depression, lethargy, and a crappy sleep pattern. When you couple all of that with not being able to sleep properly in microgravity, it's hardly a surprise to discover that 75% of astronauts report taking sleeping pills in space.

So while Breakaway Syndrome has never materialised with an astronaut trying to run off into the cosmos, depression from leaving friends and family behind is a very real concern. Astronauts on board the International Space Station have a confidential video call with a psychologist every two weeks, along with regular communication with friends and family. Highly successful Mars simulations such as NASA's HI-SEAS mission in Hawaii, Concordia Station in the Antarctic, and Russia's 520-day MARS-500 have also gone a long way towards helping researchers understand what makes people tick when they're isolated together and living in Mars-like environments. The biggest thing they've learned though? Both pretending to be in space and actually being in space can be boring. As in *really, really boring*. The difference on Mars, though, is that you can't just spend all of your time screwing around on Facebook.

Sociopathic Social Media
You might think Facebook and social media would be the perfect way to stay connected to loved ones when you're on a cold and unloving planet 56 million kilometers away, but how many people on Facebook do you even really like? Personally, I can count all of them on one hand but maybe I'm a bad example. Even if you're *not* a ginger misanthropist who's determined to die on another planet, your brain probably isn't wired to have more than 150 emotional connections. Dunbar's Number is named after Robin Dunbar, who proved that the size of a primate's brain (specifically the neo-cortex) is directly related to the biggest stable group size that primates can maintain (bigger brain equals bigger bunch of apes). So when Sally

in Accounting whinges that you haven't accepted her friend request, tell her it's because you've reached your brain's quota for emotional connections. Or you could just be honest, and tell her it's because you're not interested in seeing her stupid baby photos, like I would.

Several comprehensive studies have now shown that people primarily use social media to either present the most exciting aspects of their lives, or complain about something or someone in a public forum; and unfortunately getting this 'highlight reel' of other people's lives can cause depression. Because that's what Facebook and Instagram really are: posting carefully curated photos from your boring life to make it look more interesting, while desperately hoping for external validation from people you barely know through a social interaction that requires all the thought and energy of hovering over a button long enough to decide if you should use a loveheart or if just a thumbs up is enough. So we won't have Facebook or Instagram on Mars, because if you're living like a mole-person underground on Mars and your primary connection to friends and family is filled with stories and photos of how much fun they're all having back on Earth without you, it'll be like strapping a rocket onto the sled of anyone already on the slippery downward slope of isolation-induced depression. Personally, I know I'll be fine because I'm already looking for an excuse not to attend your event or look at photos of your stupid baby. Weddings, baby showers, Christmas, your kid's birthday party with 36 screaming five year olds: all important social events I won't have to apologise for not being at because I'll be 56 million kilometers away living underground like a ginger, Martian vampire.

Even if they're not scrolling through it themselves, social media channels will still offer an incredible opportunity for the first Martian colonists to share what life on Mars is like with the rest of humanity and make a positive impact on the lives of people they've left behind forever. That orbital perspective can drastically change someone's outlook on life, potentially influencing huge sections of the world community into realising we're all on this pale blue dot together, protected by a wafer-thin atmosphere we're pumping billions of tonnes of carbon dioxide into every year. During the final servicing mission for the Hubble telescope in 2009, NASA astronaut Mike Massimino became the very first person to use twitter from space, while Canadian astronaut Colonel Chris Hadfield reached millions worldwide during his six-month command of the International Space Station by posting breathtaking photos of Earth and entertaining videos of life in space. But for all the wonders of space, the inspiring stories of the people who will first represent our species on Mars, and the amazing opportunity to share all of this and show how fragile and unique Earth is that the internet provides us with... there are still plenty of people who'll use it to call the rest of us naïve snowflakes. The internet has brought us all closer, but it's also shown that heaps of the people we're now closer to are absolute pricks.

Whatever their reasoning, trolls are why the first rule of the internet is 'don't read the comments'. It doesn't matter if it's a YouTube video of a four year old dancing or an article on cancer research, the comment section will leave you doubting if our species even deserves to explore other planets, or if we should just set off all the nukes and save the rest of the solar system the trouble. Some valiant internet warriors try to fight back with witty arguments, but that breaks another rule of the internet: don't feed the trolls. It only makes them stronger. Imagine: you're not just sitting comfortably on the couch in your unicorn onesie, getting upset because strangers said bad things to you on the internet. Instead, you've just said goodbye forever to all of your friends and family because you've set out on a dangerous mission to colonise another planet, doing something that you believe will inspire our entire species to dream bigger, and some jerk is commenting on the livestream about how he wished your rocket blew up. Or, you're getting hate mail while you're on Mars through the only communications link you have to the species you thought you were serving.

In the case of Chris Hadfield, he was shielded from the viciousness of the internet during his command of the space station because he wasn't the one doing the actual posting to social media. Rather than have a space station commander with no background in social media try to work through posting things to Twitter and Youtube then try to respond to the thousands of comments, it was all being handled by his son Evan. Commander Hadfield would email photos and videos from the International Space Station to Evan on the ground who would then post them under his Dad's name, all culminating in a guitar-playing Commander Hadfield playing a beautiful cover of David Bowie's "Space Oddity" as he floated around the station in microgravity during his last days in space. As a social media specialist, Evan Hadfield was equipped to deal with the best and the worst of the internet, letting his Dad get on with his job: running a space station.

With the delicate and dangerous psychological pressure future Martian colonists will be under, they'll need to be able to implicitly trust mission control to pass on messages and news of what is happening back on Earth without exposing them directly to the raw, unfiltered internet. Luckily, this is where physics steps in and provides an extra layer of psychological protection.

Worse Than Dial-Up

Given that Mars, at it's closest, is still over 55 million kilometers away from Earth, you'd expect a spaceship to take a while to get there. What you might not expect is how long it takes *light* to travel between the two planets. Even at nearly 300,000 kilometers a second, it takes light over three minutes to cover the distance. So when you look at Mars through a telescope, you're really seeing it as it was at least three minutes ago. Since nothing can travel faster than light, that delay also applies to all communications to and from Mars, so you can forget about picking up the phone to call Mum when it's all getting too much. It'll take three minutes for her to hear your first sob, and even if she immediately responds with, "Did the big Martians pick on you again?" you'll still be weeping for six minutes before you hear it. No phones, Skype, or any other form of 'instant' communication, because everything is delayed by that same cosmic speed limit. That's not to say Martian colonists won't have communications, though. We'll just send and receive emails, text messages, tweets, photos, and pre-recorded videos instead. But don't forget that six-minute communication round-trip is only when Earth and Mars are at their very closest. On average it's closer to 12 and a half minutes each way, so it'll be nearly half an hour before you know if the entire internet is sharing your obnoxious Mars selfie.

This communications delay really throws a spanner in the works for the folks in Mission Control, who'd probably love to micro-manage what humanity's first representatives on another planet are doing with their time. Right from the start of the space age, NASA introduced a capsule communicator, or "CapCom", a single and familiar voice relaying commands from Mission Control over the radio. CapCom was almost always a fellow astronaut who knew the mission as well as the crew in space did, and was there to stand up for the folks in space when everyone else in

Mission Control might not have the same perspective on the situation. While a CapCom - and the flight controller yelling at them - played an absolutely critical role in the early years when space missions were short and scheduled down to the second, the Skylab debacle has shown that astronauts need more autonomy on longer duration missions. While the folks on the International Space Station still talk to those on the ground every day; outside of special circumstances like a spacewalk, Mission Control is much more likely to send a daily work list over email and trust that the crew will do everything they can to get it done. What terrifies people in Mission Control is that the crews on Mars will be making decisions far more independently than any other space explorers in history. Any emailed work order could take 22 minutes to arrive if it's sent when Earth and Mars are at their farthest apart, so Mission Control can't expect an instantaneous answer from the crew. There's also no way to demand they do anything anyway: if the crew don't want to do what they've been told, what's the worst Mission Control can do? Send them a more sternly-phrased email? None of the astronauts involved in the Skylab mutiny ever flew in space again, but you really can't threaten a crew on Mars with something like that... especially if the plan is that they're never coming back.

The other minor issue with sending selfies from Mars is that every 26 months Earth and Mars wind up on opposite sides of the Sun. Not does this set them nearly 400 million kilometers away from each other, but it also puts a 1.4 million kilometer wide ball-shaped uncontained nuclear fusion reactor directly between the planets, blocking all communications. To get around the Sun so that viewers back on Earth can take bets on who's going to start talking to the plants first, you'll need to launch two communication satellites: one to orbit the Sun, and the other to orbit Mars once a day, known as an areostationary satellite. The sun-orbiting satellite trails far enough behind Mars that when the Sun gets in the way of direct Earth-Mars communication, you'll still be able to bounce signals off the satellite. For someone on Mars, the areostationary satellite will look like a star that sits in exactly the same spot, constantly looking down on the colony to send and receive signals. So if our intrepid Mars colonists suddenly decided to put on a show for the cameras and turn their habitat into an underground nudist colony, that horrifying footage would first be transmitted from the Martian colony to the areostationary satellite above, be received by the areostationary satellite and transmitted to the sun-orbiting satellite, and be sent from the sun-orbiting satellite to a receiving station on Earth, where it would be viewed and likely destroyed for the good of humanity.

Traditionally we expect all of this to be done via. radio because those big transmitter dishes you see on space probes like Voyager, Cassini and New Horizons have to be doing something, right? But the issue with using radio is that it's *so damn slow*. It's incredible that we're still getting messages from Voyager 1 as it ventures past the edge of the solar system, as the effective data rate is 160 bits per second. Forget about 4G, this is about 60 times slower than the fax machine your Mum keeps in the cupboard, or 228 times slower than the dial-up modem your Dad has hidden in the attic in case the machines turn on us. The stunning photographs that Cassini sent back from around Saturn and its moons? They took *weeks* to transmit. Making someone who's used to having broadband go back to dial-up internet is a quick way to discover if they're going to eventually murder you while you're living together on Mars, but it's not great for everyone back on Earth who wants to watch it . If you want a continuous broadcast of the crew's nudist parade coming back to Earth in high definition, you'll need lasers.

Compared to radio transmitters, laser communication is practically ultra high-speed interplanetary fibre-optic broadband. The laser light itself doesn't *travel* any faster than radio because light's cosmic speed limit means that 3 - 22 minute delay still applies, but by firing a giant frickin' laser through a telescope you can transmit and receive far more data every second. NASA trialled a laser communication system on board their lunar probe LADEE in 2013 and managed to transmit at 622 megabits per second *from the Moon*. That's fast enough to download the widescreen Blu-ray versions of Interstellar and The Martian in less than 20 minutes, so you can spend the next seven months on the trip to Mars wondering which version of Matt Damon in space you're going to be when you get there.

The Slow Boat To Mars

It's obviously not just communications that are affected by these literally astronomical distances. Getting *anything* to Mars is tough. While you can travel there in just six months using current rocket technology, that only happens if you launch during a small window of two or three weeks when Earth and Mars are approaching their closest point. Miss that window and you'll have to wait over two years before you get another chance. You can still launch supplies whenever you like, but instead of six months, they'll take nearly a year to get to Mars.

Until someone develops a Star Trek transporter, you won't be getting any fresh fruit delivered from Earth or anything from a hardware store if part of your life support breaks. The fresh fruit thing isn't life-critical, but it's less than ideal for the crew's mental health. With most food on the International Space Station being rehydrated from inside a sealed plastic bag, the crispness of fresh fruit when it's only delivered once every three months by a resupply ship makes it a treasured commodity for the crew. That's all right when the trip from Earth to the Space Station is only three days, but your fresh apples are going to turn into the solar system's worst cider after a year in space. More worrying is having something break in your Martian life support system and not having the spare parts to fix it. Not only is sending spares for everything from Earth to Mars ridiculously expensive, it also quickly becomes unsustainable as you put more people on Mars and the Martian community grows to need more life support and more living space.

To get around the spare parts issue, Martian colonists will need to have their Mars surface suits, habitats, and life support systems designed to be easily repaired using spare parts that are mostly 3D-printed using local Martian resources wherever possible. With 3D-printed rocket engines being used by SpaceX and Rocket Lab already, and 3D-printed plastics and paints already being made from simulated Martian soil, spare parts 3D-printed from actual Martian soil will be critical to cutting down how much stuff we'll need sent to us from Earth. By also making the spare parts common across equipment, like using one type of 3D-printed fastener in your Mars surface suit instead of 5 different ones, then the spare part requirement is reduced further. There shouldn't be such a high demand for spare parts on Mars anyway; given Earth is 56 million kilometers away and there's at least a year between resupply, *everything* will be done a lot slower and with an even bigger focus on reducing risk, even compared to the International Space Station. Martian colonists will be there for the rest of their long and healthy lives as underground mole people, so unless it's life-threatening, no one is going to be rushing to do anything once they're there. That first crew in particular will be there a long time with no one else around to help out if something goes wrong. It will really be about staying indoors, playing it safe, and simply surviving for those first few years until more people arrive.

Plants: The Original 3D Printers

While the first crews will arrive on Mars with literally years of emergency food rations already waiting for them, that drive to become self-sustaining and completely independent from Earth resupply means they'll be trying to get their inflatable greenhouses growing food as soon as possible. Embracing their inner hipsters, the crews will be focused on growing hardy crops like quinoa and kale, which will provide both the highest nutritional value and greatest ironic pretense. With sunlight being half the strength it is on Earth, these greenhouses will be covered in dirt and lit inside by LED lighting, powered in turn by solar panels on the surface. Just like many suspicious underground plant labs on Earth, there's a good chance that the plant life inside will be grown hydroponically, being maintained by the colonists as both a food source and a past-time. Caring for plants on Mars will almost certainly play a huge role in stabilising the mood of the crew, because when you're living on a dusty red planet devoid of anything resembling moss, let alone a plant, looking after something green and alive will have an incredibly positive psychological effect on most people. Probably keep me out of the greenhouse though, because I've killed pretty much any pot plant or Bonsai tree I've ever touched.

Ultimately, the folks on Mars are aiming to achieve something Buckminster Fuller called "an astronaut's black box" - a perfectly closed, self-moderating, and self-sustaining environmental system where the only input is sunlight and the only output is heat. One simple example of this is a Tradescantia plant that's been thriving unwatered inside a 'bottle garden' since 1962. Retired electrical engineer David Latimer last watered the plant in 1972 before sealing the glass flask up as an experiment, and it's still growing inside it today. Dead leaves from the plant fall off, break down to become soil, release their stored carbon dioxide for the plant to breathe, all while the plant continuously recycles the water in the flask at the same time. Just like David Latimer's plant and the horrifying things I've found in my grandparent's pantry, researchers have been trying to create closed ecological systems for decades with some pretty uneven results. While the Soviets were sealing people up inside the BIOS experiments for up to 180 days at a time, and the European Space Agency has been running the MELiSSA initiative since 1989 to develop regenerative life support systems, the most well-known and infamous of these closed-loop environmental experiments that we've shoved people inside is Biosphere 2: a sealed habitat in the Arizona desert covering more than a hectare that attempted to recreate five different biomes found on Earth. During the two original missions that ran between 1991 and 1994, there were all kinds of incidents, including the structure's concrete sucking oxygen out of the air and making the crew hypoxic, accusations of secret carbon dioxide scrubbers being installed, the company running the habitat dissolving while people were still living inside it, crews fighting over whether to end the experiment early, and people smuggling in extra food because everyone was starving. Basically, Biosphere 2 was even scarier than the movie Biodome that Pauly Shore made about it, which is truly terrifying if you've ever seen what sometimes passed for comedy in the 90s.

Eating Bugs Instead Of Each Other

For folks who like dead animals on their pizza, Mars is going to be particularly challenging. Life as a vegan can be equally challenging too, though, especially if there's only three other people on the planet to preach to about why veganism is so morally superior to eating meat. The issue with eating meat in these closed-loop environments isn't just trying to shove a cow in a spacesuit: it's the sheer amount of food, water, and space that Daisy the Future Porterhouse needs before you can eat her. Not only would we only eat about 40% of Daisy in the first place, she'll need to eat about 39 kilograms of feed and drink over 26,000 litres of water to make your 800 gram steak. Comparatively, 800 grams of deliciously edible crickets require about 1.3 kilograms of feed, 0.8 litres of water, and 1000 pairs of unloving black eyes staring blankly at you from a tank about half the size of a beer keg.

While insects might not be a common source of food in many western cultures, it's quite normal to pick up a cricket kebab or chow down on a mealworm burrito in the rest of the world, and it's the perfect option for getting around the protein-deficiency/smug-overdose of going fully vegan on Mars. Over 2000 different species of insects are regularly consumed worldwide, from cockroaches in China to witchetty grubs in Australia to scorpions in Mexico. With a considerably higher protein content by weight than beef and a far smaller environmental and carbon footprint, crickets are the perfect solution for filling out our otherwise vegan space burrito in the cramped environment of an underground Mars colony. Bacon fans can rejoice too: strains of Japanese 'dulse' red seaweed have the same taste and texture as bacon when fried, so we'll be able to grow it right alongside our quinoa and kale. I might not be able to put spider-pig in a spacesuit and take him to Mars with me for later consumption, but I'll be able to grow a slippery spider-pig substitute on the top of my Martian fish tank instead.

While you're biting into your cricket and mealworm burrito with a couple of strips of seaweed bacon for extra flavour, it'll help knowing that the rest of the crew are eating it with you as well. Preparing and eating food together has been shown to be absolutely critical to keeping crews cohesive and balanced during long-duration Mars simulations, and experimenting with recipes together is a perfect way to add some much needed variety when you've been trapped inside a capsule for six months and start feeling like you want to open the airlock and "freshen things up". Aside from cooking, future Mars colonists will need to be trained in a multitude of weird and wonderful things. There's no doubt they'll need to be trained as engineers, doctors, and scientists; but being millions of kilometers from a dentist means they'll need to look after each other's teeth too. Astrobiologists will also want anyone on Mars to be trained in looking for clues forto fossilised life: - our entire species would take a quantum leap forward if evidence of life was ever found in the rocks of Mars. Even geologists will be desperate to learn more about how Mars formed, and the best way for them to get the data they're after is if the colonists know the difference between gneiss and schist. Also, before the Martian geology nerds freak out, both gneiss and schist are metamorphic so you probably won't find them on Mars. Also puns are a scourge on our species generally, and people who use geology puns particularly deserve to be shot into the Sun.

Little Murder on the Prairie

Over-arching all of this training, our intrepid Martian colonists need to understand isolation psychology: watching out for the tell-tale signs that they or a fellow crew member may be on the slippery slope towards depression or a psychological breakdown. All of our Earth-based Mars analogues can only *simulate* life on the red planet, so until we actually have people there we won't truly know how they'll respond in the Martian environment. While the US's 1862 Homestead Act and Canada's 1872 Dominion Lands Act both encouraged the settlement of North America's western territories, those who left relatively close-knit communities with a dream of making a fresh start in the barren isolation and stark surroundings of the prairieland were especially prone to 'Prairie Madness', either withdrawing socially and even attempting suicide, or turning to violence against the few who were around them.

But of course, no one knew about prairie madness until people started living on the prairie.

To ensure that their future Martian colonists will be emotionally ready to tackle whatever the 'Martian Prairie' might throw at them, Mars One have identified the five key characteristics that their astronaut candidates need: *resiliency*, *adaptability*, *curiosity*, *ability to trust*, and *creativity/resourcefulness*.

<u>Resiliency</u>
Future Martian colonists need to be able to roll with the punches. When the toilet breaks in the middle of the night and you're half a solar system away from a decent after-hours plumber, the last thing you need is someone who slumps in a corner and wails, "I CAN'T DO IT!". You just have to shove earplugs up your nose like Mark Watney and start stirring. Life on Mars *will* be challenging, but being resilient in the face of adversity is absolutely key to survival. But rather than always being resilient in the face of something life-threatening, for many the challenge will be showing resilience when things are incredibly boring or irritating. Thousands of people who initially applied to Mars One's 2013 astronaut program didn't have the resiliency to finish their online application, yet resilience is almost a defining feature of being an astronaut: Neil Armstrong, Buzz Aldrin, and Michael Collins were all rejected from the NASA astronaut corp multiple times before they went on to become the crew of Apollo 11. In the case of Clayton Anderson, he applied *fifteen* times before finally being accepted to start training as a NASA astronaut. *Fifteen times* he was told by NASA he wasn't good enough. Each time he took the rejection feedback and immediately went to work improving his application for the next year. Anderson saw each of those fifteen rejections as important: if he didn't fail, he wouldn't have learnt the lessons he needed for next time. Even when he was finally accepted on

his sixteenth application, he still wouldn't actually go to space for another *nine years*. Sticking with NASA through all of that takes serious commitment and resiliency.

Adaptability
Change and the unknown are two of the things humans fear the most, yet for those first colonists on Mars they'll be radically changing almost every aspect of their life as they step into an environment full of life-threatening unknowns. Sure, you can do your best to simulate the experience of life on Mars, but until you're actually there with no way back to Earth it's impossible to know for sure. The military has a saying: "No plan survives contact with the enemy", meaning no one can tell you what a situation will truly be like until you're there and in the thick of it, so once again we'll have to roll with the punches and adapt to whatever the red planet throws at us. There will be times where we'll need to adapt fast, making snap decisions and acting quickly, but most of the time life on Mars will be slower-paced with decisions being focussed on the very long term. Do we replace a water filter when it's been used for 50% of it's life cycle, then put it back in at a later date to maintain consistent water quality over the long term, or just use it until it needs to be completely replaced? Being adaptable will not only be vital to surviving on Mars, but with a crew of just four, even the jobs you'll need to do each day will require exceptional adaptability: in the morning you might be elbow deep in the life support system replacing air filters, after lunch you're in the greenhouse harvesting some leafy greens, and then you might get the whole crew together to watch a movie after dinner. The next day you're out on the surface replacing a faulty solar cell, and the same afternoon you're conducting heart monitoring experiments to determine how your body is responding to the long-term effects of 38% gravity. Adapt, evolve, survive.

Curiosity

Humans explore to discover more about our universe, and the first mission to Mars will be by far the biggest expedition to humans have ever taken to explore our the universe. Naturally, you would want curious people leading the charge to discover more about the red planet. To seek and desire to learn more about our universe is at the heart of being an explorer, and the first colonists on Mars will be our greatest explorers so far. You also don't want to go to Mars with some boring jerk who just wants to stay in the habitat all day saying, "Nah, I'm good. You go explore the craters and discover evidence of life on another planet - I'm going to stay here and talk to the plants". WHY DID YOU EVEN COME WITH US FRANK? THERE ARE LITERALLY MILLIONS OF PEOPLE WHO WISHED THEY WERE HERE. YOU COULD HAVE STAYED ON EARTH AND TALKED TO THE DAMN PLANTS!

Ability to Trust

On a seven month journey through the darkness of space, the last thing you want is to discover is that one of the crew has secretly eaten all your biscuits and sipped away the tiny alcohol ration you'd been saving for the first Christmas on Mars. Worse still, imagine *you* were the one stealing biscuits and alcohol and had to live with the guilt of lying to your crewmates, people who would already have risked their lives countless times to ensure your safety. People who you probably don't want to start borrowing your toothbrush to clean the wet waste processing system, or moving your bunk into the airlock while you're sleeping. The first people on Mars won't just have to trust each other, they'll have to be open and honest about their actions as well as be emotionally mature enough to serve the highest and best interests of the entire crew rather than just themselves. With future crews and supplies on the way, they'll also have to trust that the people on Earth have their best interests at heart too and aren't sending another half-trained biscuit and alcohol thief to ruin Christmas again.

Creativity/Resourcefulness
You're heading out onto the surface to inspect and fix a solar panel that suddenly went offline during the day and that the colony's rover couldn't fix. After hours of preparing your surface suit and planning the trip on to the surface, you're outside the hab and inspecting the wiring in the control box when you find out the solar panel is offline because a small lead frayed and broke off in an earlier dust storm. It's an easy fix with the right tools, but the wire-strippers you brought have seized in the extreme cold.
- Do you end the surface mission with the solar panel offline, leaving your habitat on half-power for another day before you do another trip onto the surface the next night...all because you didn't lubricate the wire-strippers properly?
- Do you stay on the surface and send a message back to Earth, waiting for up to an hour before you get a reply saying, "Are you kidding? Just figure it out - you're embarrassing us."?
- Or do you improvise, jamming the plastic-covered wire between the aluminium hinge of the control box cover and using it to strip the wire so you can fix the solar panel right away?

When a dust guard fell off the lunar rover during Apollo 17's first moonwalk and the unguarded wheel started throwing a dangerous amount of dust around, Eugene Cernan and Harrison Schmidt didn't climb back inside the lunar lander and head back to Earth, nor did they replace it with a spare dust guard they just had lying around. Grabbing some duct tape, clamps, and some lunar maps, they built a new dust guard, and spent another two days exploring the Moon.

Proof that duct tape can fix literally anything
[Credit: NASA]

Finding creative and resourceful solutions will be life-saving in an extreme and isolated environment like Mars, where only using the tools in the manual or phoning a friend back on Earth becomes patently ridiculous.

By selecting candidates based on these five personal qualities rather than their professional qualifications, Mars One is hoping to start with a pool of candidates who all have a solid emotional base and *then* train them in the skills they'll need, instead of taking the most technically skilled people and later trying to teach them how to play nice with others. Rather than taking the very best surgeon, engineer, scientist, and psychologist on Earth and sending them to Mars, Mars One wants to take people who are easy to live with and cross-train all of them in everything: two of them to the level of surgeons with a good knowledge of repairing the habitat's life support systems, the other two to the level of paramedics who are habitat-repair experts. Astrobiology, engineering, geology, medicine, even politics: you name it, and they'll need to have some knowledge of every aspect of it to become self-sufficient while living on Mars. You can train someone to be a doctor and an engineer in ten years, but you can't really train them to be less of a jerk, so Mars One basically wants four generalist MacGyvers who are also really great housemates. In case you're wondering, I obviously mean the 1980's mullet-wearing Richard Dean Anderson "Disarming nukes with a paperclip" MacGyver... not the 2016 reboot that I'd push into the airlock and vent into space.

What's critical to developing those five characteristics and further reducing the "jerk risk" is how well someone can self-reflect: ask themselves tough questions about who they are, how they're responding to their environment, and why what they're doing is important to them. For many astronauts already on the Space Station this takes the form of a handwritten journal or emails to family; for those on Mars it also might take the form of an audio or video diary. Hell, you might even find the best self-reflection is writing a sassy non-fiction book about how humans will change in body, mind, and soul by colonising Mars. Whatever form that self-reflection takes, it needs to be honest and ongoing. You won't be going to Mars until you know deep down why it's so important to you. You have to acknowledge that you're a biscuit- and booze-stealing maniac so you can fix the trail of crumbs and fumes you're running from.

Even with all these self-reflecting and self-improving Martian MacGyvers, there's still no sure fire way to guarantee one of them *won't* go bonkers on Mars.

Breakaway Syndrome will still haunt the dreams of flight surgeons and space psychologists trying to prepare anyone planning to live millions of kilometers from Earth without a return trip; but by developing these five characteristics and observing how candidates behave when they're inside mock Mars habitats for months at a time, we'll have far more certainty in the psychological stability of the multi-skilled people heading to Mars than there ever was for those heading to the Moon.

Visiting The End Of The Earth To Leave It

Going one-way sets Mars One's proposed mission apart from any other space mission conducted before. Even someone like Valeri Polyakov knew that after 14 months onboard space station Mir he'd eventually come back to Earth. To find something closer to the challenges that Martian colonists will face and the personal qualities they'll need, you have to go back roughly 100 years to another golden age of exploration: the early Antarctic explorers. While Amundsen may have been the first to reach the South Pole, and Scott may have died trying to beat him there, it's Ernest Shackleton's approach that future Martian colonists can learn the most from. Far from being a perfect hero, Shackleton embraced his flaws and reflected on them in well-maintained journals during his expeditions. Always in search of the next adventure, he often swung from being lionised in the press and hosted by rich patrons as the toast of the town to being near destitute after the failure of yet another get rich quick scheme. While he may have struggled at times to fit into broader Victorian-era society, Shackleton truly belonged out in the wild, discovering the unknown. Among Antarctic explorers Shackleton quickly became renowned for always placing the well-being of his crew first. When the Nimrod expedition was starving during their return to McMurdo Sound, Shackleton gave his daily ration - a single biscuit - to his struggling petty officer Frank Wild. When Wild tried to refuse it, Shackleton threatened to bury the biscuit in the snow rather than eat when someone on his crew needed it more.

By 1912 Amundsen had already won the race to the South Pole, so Shackleton set out on what he believed to be the final challenge of Antarctica: crossing the entire continent. After significant challenges securing funding for the Imperial Trans-Antarctic Expedition, Shackleton needed to find the right crew for his ship, the Endurance.

"Men wanted for hazardous journey, small wages, long months of complete darkness, constant danger, safe return doubtful, honor and recognition in case of success".

Many of you have probably seen this advert before, but I'll burst your bubble now and add that it's almost certainly not real. There's no record of it in *The Times*, where it was supposed to have been advertised, anywhere between 1785 and 1985; and the earliest anyone has been able to find it printed is in an American book from 1944, more than 20 years after Shackleton died. In all probability it was created years after the expedition to reflect what the journey ultimately involved, and probably dreamed up by someone who used the American spelling 'honor' instead of the English 'honour' like Shackleton did in everything else he ever wrote. The truth is, there was no need to advertise:

when news broke that the famous Antarctic explorer Ernest Shackleton was leading another expedition to the great white continent, over 5,000 people applied to join the crew of the Endurance.

From those thousands of applications, Shackleton selected just 26 people to join him on what is arguably the toughest and most inspiring tale of human endurance in history; a journey that never even *reached* Antarctica when Endurance became trapped in pack ice 50 kilometers from the coast. The crew worked for months to try to free the ship, but as winter returned the Endurance was frozen into place and the crew forced to live on the ice. They tried to stay cheerful as the weather worsened, improving their living quarters and hunting seals for fresh meat, but after a few weeks their ice-locked ship started to creak and groan - she was being crushed by the slowly shifting pack ice. Shackleton ordered all the supplies to be removed from the Endurance and within days it had been crushed and sunk, stranding everyone on the inhospitable pack ice of Antarctica. With no way of calling for help, Shackleton knew no one would come to save them, but they also couldn't stay where they were, as Summer would melt the very ice they were living on. More than 550 kilometers from the nearest huts and food supplies, Shackleton and his men began dragging their life boats across the ice.

Antarctica: come for the adventure, stay for the survival situation
[Credit: Getty Images]

The epic two-year tale of survival - with Shackleton leading the entire crew of the Endurance safely over the ice and across the Antarctic ocean in open top boats, before a final desperate dash over the uncharted mountains of Elephant Island - is much better told elsewhere: "Shackleton's Way" by Margot Morrell is brilliant for both Endurance's story and it's leadership lessons. For future Martian colonists there are two immediately important lessons to take away from the tale of the Endurance expedition firstly, Shackleton filled his crew with people who were not just exceptional explorers but also artists, musicians, and photographers, people who would be able to work well together *and* capture the human experience of life on the ice. The second lesson for Martian colonists to take from Shackleton, of course, is to always bring a banjo.

Martian Deliverance
Before attempting to drag their lifeboats over hundreds of kilometers of pack ice, Shackleton ordered the crew to abandon all their personal items except the food they were carrying, the clothes they were wearing, and an extra pair

of gloves and socks. Only two exceptions were made: expedition photographer Frank Hurley was allowed to choose 150 of the best photographic plates he'd taken before smashing the other *400*, and the ship's meteorologist Leonard Hussey was ordered to bring his 5 kilogram banjo which Shackleton claimed would be "vital mental medicine".

Banjo: When being trapped in Antarctica isn't bad enough
[Credit: Royal Museums Greenwich]

Frank Hurley was a highly recognised Australian photographer, so it makes sense that Shackleton would want some photographic record kept of their expedition and bid for survival. But by all accounts Hussey only knew six songs and *no one* on the crew was a world-class singer. But stranded in a place with 160 kilometer per hour winds and temperatures dropping below -80 degrees Celsius, the plucky sound of a banjo helped lift spirits and bring a little bit of warmth back to the crew's frozen tents. Likewise, musical instruments will play an incredibly important role when the first people on Mars are celebrating birthdays and special occasions, or even just for a bit of fun when things seem boring or hopeless. Today, Hussey's banjo sits on display at the National Maritime Museum in Greenwich outside London, and you can buy replicas, but personally I'd prefer to torment the rest of the crew with my ukulele. If you have the mental and emotional strength to stand hearing me sing Somewhere Over The Rainbow even once on my four-stringed hobbit guitar, then you've probably got what it takes to colonise Mars.

For all the incredible men that made up Shackleton's Endurance crew, one aspect that was sorely lacking was any kind of gender equality. Shackleton even had "three sporty girls" who wrote to him requesting to join the expedition, offering to don men's clothes "if [their] feminine garb is inconvenient". Alas, this was the early 1900's and the polar exploration boys club was even worse than the military test pilot boys club that started the space age.

Given that, to date, only around 10% of the people who've been to space have been women, and only two women have ever commanded the space station, we've still got a really long way to go before we approach anything resembling gender equality in space exploration. Hopefully that's about to change with the first people on Mars, though. Gender-balanced groups have been proven to work more effectively than those that are male or female-dominated - no matter what your sexist uncle might say about "a woman's place" - because equal representation encourages equal participation and equal emotional investment. It seems obvious, but when people care about what they do and know that they're valued, they contribute more. Not only does this apply to gender, but it also applies to age and cultural diversity too: the more diverse the group, the more perspective it can bring to solving problems. In the case of a four-person crew on Mars, the solution is simply to train two men and two women from four different cultural backgrounds to work together using different leadership styles and embrace that 'adaptability' characteristic everyone has been selected for.

What you really need is a four-person relationship. I'm not suggesting every few nights someone puts on Barry White and there's a one-third gravity Martian foursome in the greenhouse with the seaweed watching...although I guess if everyone was into it that would probably improve group dynamics too. Especially for the seaweed. Ideally, you'd want something more like a four-piece rock band, where each crew member contributes their own cross-section of skills to the colony and together they create something bigger than the sum of their parts. The bassist and drummer synchronise to provide a great rhythm and bassline, while a vocalist and lead guitarist can rally over a gorgeous melody; but it's not until you bring them *together* that things really start to work. Just like a band, a crew on Mars will need plenty of practice working and playing together to discover each other's quirks and talents. The band also needs to be resilient and adaptive in their style to remain functional if the drummer breaks a drumstick or the guitarist blows an amp: if two of the crew members are incapacitated or killed, the other two should be able to keep the colony running by themselves until another crew arrives to help out.

There's plenty of space psychologists who'd be concerned about a crew on Mars pairing off in romantic relationships, the same way you might worry that members of your favourite band sleeping together could destroy the entire group. But there's plenty of examples where bands have survived and thrived after inter-band relationships turned sour: Fleetwood Mac, The White Stripes, No Doubt, and Blondie, just to name a few. Hell, if ABBA can survive two divorces then a Mars crew can survive an inter-crew break-up, but obviously if anyone actually *plays* ABBA in the habitat I'll be opening all the airlocks. While living on Mars isn't going to be the opening scene of Barbarella *all* the time, having sexual relationships form within some of the Mars crews isn't just inevitable, the mixture of emotional vulnerability and stress-relief will provide a huge boost to their psychological health. The biggest issue of course is the potential end-product: little Martians.

Mini Martians

The first human child born on another planet might sound like a wonderfully romantic idea, but I'm more concerned about the practicalities of toilet training your little bundle of tears and vomit inside a multi-billion dollar colony on the surface of Mars. For starters, where are you going to store them? Unfortunately most children don't fit in the standard lockable storage units on spacecraft no matter how hard you try, and even Elton John knows Mars is no place to raise a kid; that's if you can even carry a kid to full-term in the first place. There's considerable research from the International Space Station showing that rats don't just struggle to conceive in a weightless environment, but even if you give them a little space-rat sex-nest to help out, they still can't carry a pregnancy to full term in microgravity. Without a constant downward pull of gravity to orientate it in the womb, a rat embryo doesn't know which way is up, down, or sideways; so instead of developing in the form we're familiar with, it attempts to grow in *every* direction at once. Likewise for plants, without some sense of which way is down, their development is stunted and chaotic.

Colonists on Mars would have 38% of Earth's gravity to provide some sense of down, but we still have no idea if that's enough for an embryo to develop properly. Humanity's ancestors first started standing up on their back legs against Earth's gravity 4.4 million years ago, so our reproductive systems have had a long time to produce little hairless apes that continue to be strong enough to do that. But will a human embryo developing in only one-third of Earth's gravity develop that strength? What if a Mars-born baby only has 38% bone density, or without gravity restraining them, they grow to nearly three times the normal size in the womb? It's possible that even a little gravity will correct the embryo defects seen in rats developing in microgravity, but until someone tries to breed rats on Mars we just won't know for sure. We'll need close to a decade of breeding animals in Mars gravity before there's enough data for anyone to even consider trying to have a human baby on Mars. Hopefully by then we'll have enough room in the habitat to just put it down one end of the greenhouse so I can avoid it, just like all the plants I'm not allowed to touch.

With your crew free to sleep together, but with so many unknowns over how a baby would develop on Mars, you also have to be concerned about "accidents". Personally I never really understood how it could be an accident to get someone pregnant, as if you somehow tripped and fell into someone else's crotch. Regardless, with the colony's medical facilities initially being designed for non-pregnant human adults surviving on limited food, water, and oxygen, the last thing you want is someone awkwardly saying, "Oopsie". Ethically, you could never force a crew to be sterilised, but a candidate who couldn't have kids would reduce the overall risk to the mission compared to someone who *could* spawn; which is just fine by me, because the last thing Mars needs is a bunch of diminutive red-haired nightmares that look like me running around touching stuff with their nasty little fingers.

As much as I'd like to be really far away when it starts to happen, to be a truly self-sustaining colony means we'll eventually have to start making little Martians too. When that happens, it'll raise all sorts of intriguing questions: are they still human, or are they now Martian? Could they ever survive a visit to Earth, and how would a Martian child view the way Earthlings behave? What rights would they have? Colonising Mars doesn't just mean figuring out how to keep people healthy and sane there - it also means asking deep philosophical questions about who we are as a species.

Why are some of us driven to leave the safety of our home planet to explore others, how do we change the planets we live on, how do they change us, and what kind of relationship do we want with the universe as our species ventures further beyond our pale blue dot and into the great unknown?

Martian Soul

Imagine you're in orbit 300 kilometers above the Earth, knowing this is the closest you'll ever be again to the planet on which you were born. Strapped in to your seat, you feel the pressure suit squeezing tight on your body, as you listen to the crew commander running through the final checks with Ground Control over the radio. You're staring out a window down at that blue-green marble where, 4.5 million years before, your ancestors took their first steps on just their hind legs; and you're thinking wistfully about how you're about to take an equally large first step in the evolution of our species.

You snap out of your revelry as you hear ground control radio through: "Mars One, you are GO for Mars-Transit burn."

The crew commander replies:"Copy Ground, we are GO for Transit burn. So long, and thanks for all the fish."

You push your helmet back against the headrest, ready for acceleration as the turbo pumps spin up. Within seconds you're being pressed back against your seat, feeling like a gorilla is standing on your chest as fuel pours into the engine bells; and with a low-frequency rumble you *feel* rather than hear the spacecraft beginning to accelerate. The g-forces aren't as jarring as they were during your launch from Earth 24 hours earlier. This is more like a long and unforgiving squeeze as the minutes continue to drag by, with that gorilla getting constantly heavier, making it progressively harder to breathe.

Just when you're not sure you can take much more, the g-forces ease as the flight computer throttles back the engines before it proudly announces that the ship has reached its target velocity for Earth-Mars transit. You unstrap from your seat, pop off your helmet and dash back to the window: Earth doesn't look any further away. In fact, it doesn't look any different at all... but you've already crossed the point of no return. Within a matter of days, Earth will have shrunk to just a pinprick of light and you will never set foot on it again. You've left behind the planet that every other human being in history was born on, lived on, and eventually died on.

Not you though, you won't die there, nor will anyone else on board the spacecraft. After fifteen years of training and preparing for life on Mars - learning everything from how to repair a robotic rover to how to cook a perfect ground-cricket burger - you've now broken free of the Earth's gravity, something only twelve others have done prior, and something no-one has done since 1972. Within 48 hours you've sailed passed the Moon, going further from Earth than any member of your species before you and headed for a final destination hundreds of times further still. Looking out into the empty darkness of space with just the laws of gravitation now guiding you, knowing there's very little to do besides waiting out the next seven months until a cold and hostile red planet looms in the window...what would you be thinking?

I know exactly what I'd be thinking: "This is bloody awful."

Honestly, imagine just being stuck in a tiny metal box being irradiated for the next 210 days, eating unheated food out of plastic army ration bags, not being able to wash properly, and having to wake up every morning to stare at the same stupid three faces all so you can land on a cold, desolate planet with no hope of ever returning to friends and family. Not only that, but your every move will be recorded and transmitted back to Earth when you finally land: every mistake hyper-analysed, played back in slow-motion, broken-down, and eventually re-mixed into some idiot's music video. You will literally have an entire planet of people watching everything you do, or at the very least catching the gag reel on the 6pm news each night.

Just getting off the ground is damn dangerous. Of the roughly 600 people who have been to space, 18 of them have been killed during their missions Vladimir Komarov on board Soyuz 1 when the parachutes failed to deploy during landing, three cosmonauts on Soyuz 11 when it depressurised, seven astronauts on Challenger when it disintegrated during launch, and seven on Columbia when it disintegrated during reentry in 2003. That's not even counting those killed in training accidents just preparing for spaceflight. It's not like our history of sending *anything* to Mars is particularly good either: of 46 spacecraft sent to Mars, 27 have failed. Admittedly, 18 of those 27 failures were spacecraft sent by the Soviet Union, who were really just chucking stuff around the solar system as fast as they could in the 70s and 80s. Still, the stats for landing stuff on Mars aren't great. Even if you *are* successful, what a monumental waste of money sending people to other planets is when we have such extraordinary problems here on Earth to solve first! How many starving people in Sudan could you feed with the money a Mars mission would cost? How many refugees could be helped in war-torn Syria? What kind of monsters would sign up for a one-way mission in the first place? How incredibly selfish to leave friends and family behind on Earth all for some adventure on Mars where there's a good chance you'll die before you even get there, right? What could possibly drive these intelligent and capable people to risk life and limb by literally riding rockets into space to a desolate planet they can never come back from?

Call to Adventure
One of the most common arguments for exploring space is also one of the most backward: national pride. Neil Armstrong and Buzz Aldrin may have left a plaque on the

moon reading, "We came in peace for all mankind", but there's a reason we call it the "Space Race". If the US hadn't been so terrified of the Soviets beating them there, would they have have bothered going to the moon at all? Nationalism might have gotten the US to the Moon - and by extension humanity as well - but it didn't get them any further. Plenty of people might talk about how the US "won" the space race, but what did they really *win*? Richard Nixon radically reduced NASA's funding after Apollo 11, the entire Soviet Union collapsed, and no one has sent people back to the Moon since 1972. As Eugene Cernan, the last man on the Moon, said in a stirring speech before Apollo 17 lifted off from the Moon for the last time:

"As I take man's last step from the surface, back home for some time to come (but we believe not too long into the future), I'd like to just say what I believe history will record: that America's challenge of today has forged man's destiny of tomorrow. And, as we leave the Moon at Taurus-Littrow, we leave as we came and, God willing, as we shall return: with peace and hope for all mankind."

Cernan also remembers saying "Let's get this mother out of here!" after his speech, but it's curiously absent from NASA's official mission transcripts.

That's not to say that NASA didn't have plans beyond 1972, they just started cancelling them when they knew they were going to beat the Soviets to the Moon. Right around the time Neil Armstrong and Buzz Aldrin were becoming the first men to walk on the Moon, NASA was already cancelling Apollo 20 and repurposing its already-built rocket for launching the Skylab orbital laboratory instead. Apollo 18 and 19 would have taken six more men to the Moon in 1973 and 1974, but were both cancelled after the Apollo 13 disaster. The Apollo Applications Program was intended to follow on from the successes of Apollo to take the US further still, with plans for 6 men living in a Moonbase for up to 6 months, a crewed fly-by of Venus in a Skylab-type habitat, and even nuclear-powered rockets to power American astronauts out towards Mars and beyond. In 1948, long before he designed and developed the US's Saturn V lunar rocket, Wernher von Braun published "Das Marsprojekt" as the first detailed study of putting humans on Mars, with the aim of landing 70 people on Mars by 1965. Von Braun envisioned a constellation of spacecraft travelling to Mars, with one craft landing on Mars's north pole using specially designed skis, allowing the crew to trek 3000 miles down to the equator and build a runway the other nine spacecraft could land on. Just constructing the spacecraft in orbit was expected to require 1000 3-stage rocket launches from Earth to get all the components into orbit. Considering the entire Apollo program cost a little over $25 billion, this was obviously going to add up, so Von Braun later revised the plan to launch just three spacecraft to Mars, which would require *only* 400 launches.

Von Braun continued to advocate for a human mission to Mars within the next ten years right up until his death in 1977, but once the US landed on the Moon they no longer needed to prove they could. With the lunar program requiring 4% of the US's GDP to maintain, there were many who thought the money could be far better spent on the war in Vietnam. So rather than follow Von Braun and the Apollo Applications Program pathway out into the solar system, NASA took the "safe" route and built a bus to space. The space shuttle was seen as the first step in turning space flight from the extraordinary to commonplace, but in many ways it also represented the US giving up on bravely sending people out to explore the solar system, and exchanging it for staying closer to Earth to launch satellites, which is far safer and more profitable than venturing into the unknown.

There has never been any shortage of extraordinary and motivated individuals pushing humanity to explore beyond the safety of our blue-green marble, but you need something more than national pride to justify the cost. The real issue with reaching Mars for the glory of a specific country is that, to make it worthwhile, you have to bring your astronauts home again to rub it in everyone else's face. It might be all well and good having the first American or Russian on Mars, but why would they stay if they love their country on Earth so much? How can they really serve as an example of that country's greatness unless they eventually return to it and tell people from their specific patch of Earth that their patch is better than the rest?

When your motivations for exploring space are solely for the greatness of a single country, it's probably better to have people working towards the *dream* of going to Mars than actually *going* to Mars. Even with the cancellation of extended lunar missions, after the first Moon landing, the crew of Apollo 11 all spoke about getting to Mars in the early 1980's. In the 1980's, many public discussions were encouraging NASA to put people on Mars by the year 2000. The documentary I watched with my ginger sky unicorn floating around in microgravity was made in 2000, and included a NASA-JPL engineer talking about how, "We will be ready for a mission to Mars in ten to fifteen years' time". When we're talking about humans *colonising* Mars, though, the idea that people from one arbitrary patch of dirt on Earth are somehow better than people from another patch doesn't really add up. If you're living on Mars, at what point do you stop being from one part of a planet 56 million kilometers away, and start just being a human that lives on Mars? When you think about humans becoming a dual-planet species, the concept of being attached to a particular nationality starts to feel a little like astrology. Where you were born on Earth shouldn't have any more influence on your worth as a human being than the apparent position of stars during a particular month should.

Planetary Perspective

While the US and Soviet governments were measuring the size of their rockets and racing to plant a flag on the Moon, something strange was happening to the people they had sent to space. In the middle of the Cold War, these people - who were supposed to be the shining stars of their nation - were suddenly talking about how we are all human regardless of nationality, how beautiful our Earth is, and how we have to overcome our petty differences to work together and protect this planet. It started with the very first man to orbit the Earth:

"Orbiting Earth in the spaceship, I saw how beautiful our planet is. People, let us preserve and increase this beauty, not destroy it!"

Yuri Gagarin's comment could easily be written off as a left-wing pinko communist ploy to destroy the American economy, steal your conservative wives, and turn your children into radical greenie environmentalist hippies; except when World War II sailor, fighter pilot, and all-American hero Alan Shepard became the first American in space, he came back to Earth and said something similar:

"I realized up there that our planet is not infinite. It's fragile. That may not be obvious to a lot of folks, and it's tough that people are fighting each other here on Earth instead of trying to get together and live on this planet. We look pretty vulnerable in the darkness of space."

Maybe Alan Shepard had been hiding his Communist Party card all along, but then Neil Armstrong walked on the Moon and *he* somehow turned into a hippie too:

"It suddenly struck me that that tiny pea, pretty and blue, was the Earth. I put up my thumb and shut one eye, and my thumb blotted out the planet Earth. I didn't feel like a giant. I felt very, very small."

Worse still, Michael Collins stayed in orbit around the Moon during Apollo 11 and came out with an even *more* outrageous attack on all that we hold dear:

"I really believe that if the political leaders of the world could see their planet from a distance of, let's say, 100,000 miles, their outlook would be fundamentally changed . The all-important border would be invisible, that noisy argument suddenly silenced. The tiny globe would continue to turn, serenely ignoring its subdivisions, presenting a unified facade that would cry out for unified understanding, for homogeneous treatment. The earth must become as it appears: blue and white, not capitalist or communist; blue and white, not rich or poor; blue and white, not envious or envied."

Although, I suppose being trapped in a little tin can orbiting the Moon by yourself for a day might make you say some weird things. The problem is as the decades have rolled on and people have kept going into space, they've kept coming back to Earth talking about how, from low-Earth orbit, there are no borders: there is only the stunning majesty of the Earth's mountain ranges, the vast stretches of the Pacific and Atlantic Oceans, the amazing power of cyclones and lighting viewed from above, and the dancing lights of the aurora on top of our whisper-thin atmosphere. Many astronauts describe a spiritual experience rather than a religious one; instead of "meeting God", they feel an intense connection to the Earth, a recognition of just how fragile and beautiful it really is, and how we so often take for granted just how lucky we are to exist at all.

For those who went to the Moon, this sense of planetary connection is even more pronounced. Rather than viewing the Earth as an ever-present friend like the crews on the International Space Station do, from more than 384,000 kilometers away the Apollo astronauts could see Earth as a whole, experiencing an undeniable connection to that small blue-green orb floating against the velvety darkness of space and instantly recognising the ridiculousness of national politics. Apollo 14 astronaut Edgar Mitchell summed it up perfectly after his moonwalk:

"You develop an instant global consciousness, a people orientation, an intense dissatisfaction with the state of the world, and a compulsion to do something about it. From out there on the moon, international politics look so petty. You want to grab a politician by the scruff of the neck and drag him a quarter of a million miles out and say 'Look at that, you son of a bitch.' "

With that kind of perspective, it's difficult to use the "we should fix the problem on Earth before we explore other planets" argument I hear thrown at space exploration so often. There's no doubt that humanity has its fair share of problems to solve, but from an orbital or lunar perspective not only do a lot of the things we argue so viciously over seem insignificant, they also seem much easier to deal with. Why go to space when there are children starving in Africa? Because the research that goes into developing nutrient-rich food that can be stored for years in space without refrigeration also provides solutions for people on Earth in regions affected by famine. Why examine the atmosphere of Venus? Besides a general desire to discover more about the universe we live in, studying Venus also provided the first example of a runaway greenhouse effect and kick-started research into climate change on Earth. There are so many unexpected spin-offs from space exploration, but no matter how amazing the technology people are still going to complain about 'wasting money on space' in a Facebook post, probably while using a smartphone that runs on silicon chips first developed for the guidance computers in the Apollo command and lunar modules. Just like national pride, there's never going to be enough new technology developed to justify a mission to Mars. But you know what *can* justify a mission to Mars *and* share the overview effect with all of humanity? Really, really great TV.

Primetime Spacetime
My parents tell stories of where they were when Apollo 11 landed on the Moon: with fifty kids crammed into the school principal's living room because he was the only person in the district who could afford a black and white television. The US had more PhD applications in the ten years after the Apollo program than any other time in history. By witnessing humankind's first footsteps on the Moon, millions of people were inspired to study science and engineering in the hope of helping make humanity's *next* giant leap a reality. Conversely, the closest thing my generation has to that kind of, 'where were you when...' experience is watching live as a passenger jet slammed into the south tower of the World Trade Center while the north tower was already burning.

Instead of looking at the Earth from 300 kilometers up and waiting for that one-way boost to Mars, imagine instead that you're watching the livestream from the spaceship. Maybe you were bitching on Facebook earlier about how it was a massive waste of money, but now you and your friends are gathered around someone's phone watching the first humans power away into deep space on a seven-month journey to permanently colonise another planet. Imagine broadcasting an event so momentous that it redefines who we are as a species, something that people will remember where they were when it happened for the rest of their lives. Imagine calling your kids into the living room to watch the first person set foot on Mars. What kind of positive global impact could an event like that have today, especially when *billions* of people can watch high-definition cat videos from a device that fits in their pocket? Even better, imagine a generation of kids growing up in a world where people already live on Mars: they won't go to school and learn that one country *used* to go to the Moon back when their grandparents were kids, they'll go to school and learn about how humanity is now a *dual-planet species*. A generation of kids growing up learning that nationality, gender, sexuality, and education aren't barriers to exploring space and colonising other planets; being taught instead that you just need dedication and the right blend of personal qualities. The first colonists may never come back to Earth, but every step of humanity's unfolding story will still be shared across the darkness of space using a video camera and a communication satellite.

Which is, of course, why it's all being turned into a reality television show...

Big Brother - Martian Edition
Reality television generally likes to inject extra drama to make things seem more exciting than they really are like when a farmer is choosing a wife while trying to lose fifteen kilos by starving on a deserted island and he has to marry her before the first time they meet in person, and then to add some excitement the groomsmen have to fight a bunch of ninjas before the wedding...probably while the ninjas are

on fire, too. On Mars, though, you won't need to fabricate a fight or make half the colonists wash the other's clothes as part of a 'Weekly Challenge' to spice things up for ratings: you'll be too busy making sure the life support system doesn't break, the habitat doesn't decompress causing you to evaporate into a wispy pink mist, and keeping everything clean so some mutant Martian bacteria doesn't start turning everyone into zombies or something. So instead of watching fabricated brain rot, the people of Earth will be watching as four humans from across the globe work together to get over their differences and start a new life on another world; not separated by arbitrary national borders, but living together in a colony on another planet.

With Great Oranges Comes Great Responsibility

With an audience of billions watching your Martian life, you'll need to be aware that the cameras are watching every time you make a bad joke about setting fire to a ninja, flirt with a crewmate as you suck breakfast out of a plastic bag, or fart in the kitchen and blame the greenhouse. Every move, every sideways glance, wink, or weird look will be analysed, re-analysed, viewed from a different camera angle, rewound, remixed, and rematched. Then, billions of your hairless ape kind are going to start throwing metaphoric peanuts at you from their electronic viewing gallery 56 million kilometers away. The first people on Mars will be ambassadors for our entire species, and with the entire planet watching, you just *know* some jerk is going to remix that weird half-second noise you made last week into a viral music video starring a bunch of dancing llamas.

John Young is the perfect template for what happens when an especially "human" moment gets transmitted to the folks listening at home. The crew of Apollo 16 had been given orange juice to drink on their mission. A *lot* of orange juice. So much orange juice that Young made a comment on the way back from the moon that he would, "Never drink orange juice again" and they would, "Bury me in oranges", all while farting uncontrollably. It all sounds like pretty harmless banter between three guys in a tiny capsule 384,500 kilometers away orbiting the Moon, except the microphones had been accidentally left on so anyone with a receiver could hear it. With NASA keeping radio communications open for any and all to listen in, everyone from the astronaut's wives, the newspapers, and the television media were listening in to every transmission to and from the spacecraft. It also happened that Florida was best known for two things in the early 1970's: launching men into space from Cape Canaveral, and growing oranges for orange juice. The governor of Florida had to make an official statement claiming that John Young wouldn't have felt that way if he were drinking *real* orange juice instead of NASA's flavoured Tang, but orange sales still plummeted and the damage was done. John Young wasn't even the one that NASA was concerned about, either: Charlie Duke had to undergo hypnosis before the launch because he notoriously swore like a sailor.

There's a huge level of vulnerability associated with being filmed and broadcast 24 hours a day, especially knowing billions of people on Earth are watching and judging you. So should our Martian colonists be filmed in the first place? Do these human ambassadors to another planet deserve their privacy, do they lose that when they become representatives for our species, or is there some mid-point in-between? Mars analogs such as Hi-SEAS in Hawaii and the Mars Desert Research Station in Utah don't place cameras in their simulations out of concern that those inside will modify their behaviour when they know they're on camera. Mars colonists won't have the luxury of hiding their farts or their sailor-mouth though, any more than they'll be able to hide potential mental health issues. With such a heavy focus on the psychological aspects of a one-way colonisation mission, anyone putting their hand up to go will need to be completely open with both the selectors and the billions of TV viewers about every aspect of their lives and personalities. There can't be a Jekyll and Hyde-type change when the camera is on. People on Mars can't afford to put on a face and hide their faults and insecurities; they'll need to be open, brave, and vulnerable about their strengths *and* weaknesses. They'll need to know themselves incredibly well and be honest with everyone they meet about every aspect of themselves: you simply can't afford to send humans to Mars and then only truly discover who they are once they get there.

This kind of openness and honesty in astronauts is really only possible when you know you're leaving Earth and not coming back. On Mars you might have the attention of billions of Earthlings on you, but it'll also be really hard for you to give a damn what anyone on a planet 56 million kilometers away thinks unless it's critical to your survival. With at least a six minute communication delay and seven months for anyone to show up in person and tell you that Earth thinks you're a jerk, the psychological pressure to live up to the expectations of anyone except the rest of the crew will be practically non-existent for anyone on Mars. You'll still want to select people who represent the best qualities of humanity as ambassadors to another planet, but you also won't have to worry about journalists chasing them around the habitat trying to get a quote about some news story that will blow over tomorrow. Don't like me farting and swearing about having to eat another kale smoothie? Then call the cops, I don't even care.

One-Way Freedom
While the first people on Mars would be lauded as brave and fearless heroes back on Earth, if they go one-way they'll also never have to deal with all the stress that's associated with Earthly fame either. While Neil Armstrong

may have been an incredibly humble and slightly reserved man, he was regularly labelled a "recluse" by journalists that couldn't get an interview with him. He still took part in hundreds of interviews but didn't seek the spotlight, and once he left NASA he became an Aerospace Engineering professor at the University of Cincinnati, inspiring the spacecraft designers and potential astronauts of the future. At the same time, Buzz Aldrin had to deal very publicly with his alcoholism, as well as being regularly reminded he was second on the Moon. It's about perspective: Buzz may have been second to step on the Moon, but he was also *walking on the Moon*! It's like having people turn around and say, "But you'll *die* on Mars!" and I yell back "Yeah, and you're going to *die* on Earth, like everyone else in history. I'm going to die *on Mars!*". If being *first* is really that important, Buzz had several firsts that he has every right to be proud of. That classic photo of a footprint in the lunar dust? That's Buzz's boot-print. He took the photo of it shortly after stepping on to the moon as a record of the softness of the lunar dust. That world famous full-length photo of an astronaut standing on the moon looking at the camera with the visor pulled down? That's not Neil Armstrong, that's a photo he took of Buzz. Of all the 37 photos taken on the surface of the Moon during Apollo 11, there's only two that include Neil Armstrong: one of Armstrong's back as he started to climb back into the lunar module at the end of their moonwalk, and the other a reflection in Buzz's helmet from that full-length photo he took. Every other picture from Apollo 11 of an astronaut on the moon is Buzz Aldrin. Buzz can also lay claim to another lunar first: during the two hours between landing and conducting the moonwalk he also became the first man to pee on the Moon.

By not coming back to Earth, you won't have to constantly justify your Earthly existence, be questioned by journalists about the same single thing from your past, or be trotted out in front of the cameras to promote something like air freshener. "There's nothing worse than farting in your spacesuit, but with Fragricent™ our thermally-cooling underwear always stayed lemon-fresh on Mars!". One of the greatest challenges after Apollo 11 for Buzz Aldrin was being regularly asked one remarkably simple question: What did the Moon *feel* like? As "Dr. Rendezvous", he'd earned his PhD calculating how to conduct rendezvous and docking maneuvers that were critical to the success of the Gemini and Apollo programs. But while aerospace engineering might get you to the Moon, it can't help you describe what standing on it *feels* like. In Aldrin's own words, "We need to have people up there who can communicate what it feels like, not just pilots and engineers". Martian colonists might be protected from the pressure of returning to Earth, but they'll need to be more than just enthusiastic and knowledgeable about surviving on Mars. As ambassadors to all of humanity, they'll still need need to communicate their discoveries, share their triumphs and their sorrows, and bring all of humanity along on their journey. As Buzz says, the first people on Mars will need to be able to tell the rest of us what living on another planet *feels* like.

Being able to relay that human experience is also a huge reason to send humans to Mars rather than simply more robots. While everyone might love to anthropomorphise rovers like Curiosity and talk about how cute it is taking selfies on Mars, it's still just a tool for *human* exploration. Robots don't need to be fed, don't get tired, and don't have to leave loved ones behind, but they also can't tell us what Mars feels like. It's surprising how regularly I have to remind people that robotic probes and rovers are not people: no one threw a ticker tape parade for Japan's Hayabusa probe when it returned to Earth after visiting an asteroid. Likewise, the Curiosity rover doesn't have a personality, it's a 1-ton, nuclear-powered, laser-firing robot that has a bunch of humans controlling it and making it draw giant red dicks on Mars.

BEHOLD, MY MIGHTY MARTIAN SCIENCE PHALLUS!
[Credit: NASA]

Curiosity leaving obscene track marks might be amusing now, but why would *anyone* care about a photo from a robot once there are *people* on Mars to take photos instead? The photos of the lunar lander from Apollo 11 are beautiful, but we're more likely to connect with the guys in those white spacesuits in each photo than the empty and desolate scene their story is set on.

Besides not being able to share a human experience, it also doesn't help that doing anything robotically on Mars is *so slow*. Robots are designed to do a few things really well, but they can't adapt to changing situations the way humans do. Every move the Curiosity rover makes needs to first be planned on Earth and transmitted with that 3 - 22 minute delay before the rover can blindly carry out the order, taking a photo or a measurement, and sending that data back across the blackness of space for humans to interpret before they plan the next step. So while it's amazing that we have something the size of a small car driving around on another planet, that huge communication delay and lack of adaptability means the Curiosity rover did less science in its initial two-year mission than what a trained geologist could have done in a single week by physically turning over rocks on Mars. While robots and rovers will improve over time, they're still incredibly slow and we don't fully trust the results they send us. The reality is, if we're ever going to find conclusive evidence of past or present life on Mars, we're going to have to send people there.

Commando Critters

For hundreds of years we've envisioned life on Mars, human or otherwise. There's no doubt that putting humans on Mars would change the way we see ourselves as a species, but the prospect of finding evidence of life on another planet would radically change our outlook on the universe. Given that Mars is a freezing desolate hellscape, we're probably not going to find Marvin the Martian lurking there in an underground lair. But we've already found lifeforms here on Earth that could survive being directly exposed to the Martian environment, and would likely thrive in the lava tubes and caves on Mars where conditions aren't quite as fierce. Tardigrades are microscopic creatures shown to survive in places that we've never imagined even hardy bacterial life could withstand. We've exposed them to the vacuum of space, irradiated them with more than a thousand times the lethal dose of radiation for humans, and frozen them to minus 272 degrees Celsius just to name a few, and yet they keep surviving. These incredible creatures are one of the few species to have survived more than half a billion years, through all five of Earth's mass extinction events, and they're basically the animal kingdom's miniature version of Rambo.

A live tardigrade casually chilling out in a 30,000 Volt electron microscope beam
[Credit: <u>Daniel Mietchen</u>]

Even if we didn't find something as tough as a tardigrade on Mars, there are still plenty of examples of utterly terrifying microbes and bacteria on Earth that are even tougher than tardigrades. Microbes on board the International Space Station have been reported to adapt to space conditions to become even more dangerous in space, while researchers in Germany have shown that a species of lichen exposed to Mars-like conditions for 34 days didn't just survive, it started to *adapt* to it. If the first Martian colonists find something like *that* hiding under a rock, you can be sure that everyone back on Earth will be glad it was a one-way trip. In the event that there is some kind of space-Ebola waiting to leap into the bodies of Mars colonists, at least the folks on Mars will be the only ones dying instead of bringing it back to Earth to wipe out small towns in Arizona à la *The Andromeda Strain*.

Keeping everyone and everything on Mars provides a nifty quarantine system that the Apollo program didn't have. When Neil Armstrong, Buzz Aldrin, and Michael Collins returned from the Moon aboard Apollo 11, they donned hazmat suits before getting into the helicopter, were placed into a specially-made quarantine trailer as soon as they were aboard the USS Hornet, and had to stay there for three weeks to make sure they hadn't brought anything hideous back from the Moon. Apollo 13 avoided quarantine because they were too busy trying not to die from their spacecraft exploding to worry about landing on the Moon in the first place, but the crews of Apollo 12 and 14 had to go through the same three week indignity of being zoo exhibits on an aircraft carrier. A year after the first Moon landing, NASA decided lunar dust probably didn't have anything awful in it, so the crews of Apollo 15, 16, and 17 were allowed to start shaking hands and kissing babies as soon as they returned to Earth.

We might avoid spreading Martian plague on Earth by staying on the red planet to study it, but if we're going to go looking for life on Mars it's also important to remember that humans are disgusting skin bags of mucus, germs, and bacteria too. The average human mouth contains between 500 and 1000 different types of bacteria, so if we're going to find *Martian* bacteria in a rock it'd be really great if no one spits on it to clean away the dirt first. Even without anyone spit-shining the rocks we're still pretty revolting creatures, so our first colonists will be wanting to set up their underground home a *long* way from anywhere we might expect to find life. One of the best candidate locations for a future Martian colony is the sparse, open plains near where NASA's Viking 2 lander touched down in 1976 on Utopia Planitia, hundreds of kilometers from anything even *resembling* a lava tube, cave, or hill that might have sea monkey-filled brine water seeping out of it.

Viking 2 soaking up the serenity in a Martian boulder field
[Credit: NASA]

From our quarantined human colony on Utopia Plantia we'll be able to remotely control sterilised rovers in *real time*, sending them out to look for evidence of life in those protected sites without humans contaminating them, and all without the communications delay of trying to drive them from Earth.

Living Off The Land
Besides helping to keep our filthy paws away from anything that might be alive on Mars; Utopia Planitia is also an ancient flood plain with plenty of flat ground to make landings easier, and with vast quantities of water frozen in the soil to dig up and process into drinking water and oxygen. With engineering and financial limits on what can be sent from Earth, this 'In-Situ Resource Utilisation' is ultimately what is going to make any Martian colony sustainable. Colonists will need to adapt and build things using as much recycled material as possible, while using Martian soil and the thin atmosphere to provide the essentials for life like water, atmosphere, food, and shelter. Rovers can scoop up the hydrated soil of Utopia Planitia, dump it into an oven on the life support system to extract

the water, then take the now-dry soil and deposit it over the living habitat to provide extra protection from radiation. Extracting oxygen from some of the water could help pressurise the habitat, along with nitrogen and argon found in the Martian atmosphere; while hydrogen from the water could be reacted with carbon dioxide found in the Martian atmosphere to produce methane, a perfect fuel for powering Mars rovers, and can also be used for heating and power during the long Martian winters. Researchers are already manufacturing bricks and 3D printing tools using Mars soil simulant, so instead of sending countless spare parts from Earth, Mars colonists will be able to simply print their own in new Mars-brick habitats.

Conserve or Consume?
Using a few Martian resources to keep the first people alive on Mars might be all well and good, but as more people arrive we need to ask ourselves what kind of relationship we want to have with the red planet. Do we try to keep it as pristine as possible to preserve its stark natural beauty, or do we mine it relentlessly and build a fleet of starships to explore the galaxy? One of the ideas regularly thrown around in science fiction is terraforming: manipulating a planet to make it habitable for humans. You probably saw it at the end of Star Trek II: The Wrath of Khan, with - spoilers - Spock's space-coffin landing on a planet that had been terraformed into a lush tropical paradise. I didn't see it though, because I was nine and still huddled in a corner convinced some horrifying ear bug thing was going to pop up, chew its way into my brain, and force me to murder Captain Kirk.

Rather than firing a giant torpedo that blows up a planet then rearranges it into something habitable from the subatomic level up, a slightly less explosive way to make Mars more habitable for humans is to orbit huge mirrors over the frozen poles of Mars to melt them. While melting the poles on *Earth* is generally accepted to be a bad idea by anyone who doesn't own a coal mine, melting the water and carbon dioxide ice on the poles of Mars will slowly warm the whole planet and thicken the atmosphere to the point where you could walk around on the surface without a spacesuit on. It might take hundreds or even thousands of years to do it, and you'd still have to carry your own air supply because breathing the atmosphere would still be like sucking on a car exhaust, but at least the carbon dioxide-heavy atmosphere would provide a little extra radiation protection, and you wouldn't need to worry anymore about every fluid in your body boiling if someone cracks open a door outside.

Other suggestions for warming up Mars have included slamming an asteroid loaded with ammonia into it: provided it's not so big you crack the planet's core, the ammonia will disperse over Mars and work as a highly effective greenhouse gas to help melt the poles in place of mirrors. Elon Musk has even talked about dropping hundreds of thermonuclear weapons over the poles, although I doubt any astrobiologists appreciated Elon's sense of humour. That's the problem with terraforming: just like the Star Trek Genesis device, anything you do to change Mars from the dry and frigid planet we see today into something warm, wet and inviting will destroy the planet's naturally desolate yet pristine environment. With a wisp of an atmosphere and little, if any, geological activity for the last four billion years, Mars is practically a museum. Any fossil on Mars would be incredibly well preserved, but the moment Mars has an atmosphere and a little more heat, it'll also start to have weather again: weather that will cause erosion and likely wipe out any chance of finding fossils of ancient Martian life. The other danger with terraforming is that as soon as you make Mars more habitable, you'll also be making it possible for anything that might be lurking in the soil to spring back to life. Worse still, any bacteria you bring with you could mutate, and besides consuming any evidence of Martian bacteria, it might spread over the entire planet. You might spend a thousand years warming Mars so you can walk around without a spacesuit, and then not be able to even *go* outside because some mutant strain of Chlamydia that eats spacesuits and flesh has taken over the planet: another reason to get your koala checked before you take it with you to Mars.

Call The Cops, I Don't Even Care
To make sure no one starts launching nukes or STI-riddled marsupials into space, the United Nations created the Outer Space Treaty in 1967. Designed to ensure that space is used only for peaceful purposes, it expressly prohibits anyone from putting weapons of mass destruction into Earth orbit, or on any celestial body. So strictly no nuclear weapons on the Moon, which is probably just as well considering the US had been secretly planning to nuke the Moon in 1959 to prove to those pesky Soviets they could. I'm not even joking - the US was going to nuke the Moon. It was called "Project A119". Go look it up. Article IX of the Outer Space Treaty also adds provision for protecting other planets from whatever hideous things we might send at the Moon and other planets, as well as making sure we don't bring space-Ebola back. These are all pretty useful things for 87 different countries to agree to if we're going to start exploring the solar system as a *species,* rather than as *nationalist jerks.*

Where things get a little more complex for our intrepid Martian colonists is the provision that whichever country a spacecraft launches from has to remain responsible for it, and that no country can lay claim to the Moon, or Mars or anything else. At first glance this sounds great, but what about when you want to start digging up the soil on Mars to produce water and oxygen for your colonists? Under the treaty, your habitat is technically the sovereign property of the country you launched it from, but no one can claim the soil it's sitting on: that belongs to *all* of humanity. So are you allowed to use it keep your colonists alive? If you are, what happens when your colony has grown into an entire city of people, all of whom you need to keep alive? The United Nation's Office Of Outer Space Affairs (UNOOSA) has been tussling with questions like this for years, and so far no one has come up with any solid legal decisions because no one has actually colonised another planet yet to provide them with a test case. There's also the little issue of enforceability: it's pretty hard to punish people for breaking the rules when they're living seven months away across 56 million kilometers of interplanetary space. One realistic option for administering Mars, however, is to copy Antarctica. While different countries are responsible for protecting specific regions of Antarctica, none of them can lay claim to it and are prevented from exploiting the natural resources in their region. So rather than doing anything drastic like heating, nuking, or slamming asteroids into Mars, it could be protected and studied the same way Antarctica is, by small groups trying to make the lowest impact on the environment. We could keep the number of colonists on Mars small, only sending people to study and understand this desolate yet beautiful planet. Rather than 'ruining Mars just like we're ruining Earth', as every idiot on Facebook screams at me whenever I post something about the red planet, we could instead keep it as pristine as possible and conduct scientific research.

If you want to ruin anything, go and wreck the asteroid belt. No one gives a damn about that, just as long as you don't knock some massive rock onto a collision course with Earth that Bruce Willis and his oil-drilling buddies have to blow up. While many futurists believe that whoever makes near-Earth asteroid mining profitable will become the world's first trillionaire, the solar system's most substantial asteroid belt sits further out between Mars and Jupiter. So instead of trying to latch onto one of the stray asteroids that occasionally wanders close to Earth and maneuver it into orbit around Earth or the Moon, dragging asteroids into orbit around Mars and mining them there will be considerably simpler and more profitable. If we're ever going to build the truly epic spaceships of science fiction, we'll need to build them *in space* where they don't have to fight through a planet's atmosphere or strain against the pull of gravity, and to have enough resources in space to do that, we'll need to mine asteroids. If we managed to build a small rocket factory on Mars, the reduced gravity will make getting into orbit will be a lot easier, so tiny Soyuz-like spacecraft could launch humans into orbit around Mars where they could then dock with the larger spacecraft being built in space around the planet. Reducing the impact on Mars as much as possible, while utilising the vast wealth of resources to be found in the asteroid belt sounds like the best of both worlds.

Carry-On Only
If you've gotten as far as mining asteroids and building spaceships in orbit around Mars, you could almost certainly come back to Earth too. But would you want to? Physics professors regularly like to torment their undergraduate students with the Twin Paradox, a thought experiment meant to demonstrate the weird twisting of time that special relativity predicts. Imagine having identical twins on Earth, but one of them launches to Proxima Centauri b: the closest planet outside our solar system. If the twin travels at 70 or 80% of the speed of light to Proxima Centauri b and then immediately turns around again, because of time dilation, they will have aged about six or seven years while their twin still living on Earth could have aged ten. The question I had when I first heard this wasn't, "Gee, special relativity is really messed up, but isn't physics weird and wonderful?". It was, "If you had a spacecraft capable of

travelling 70 or 80% of the speed of light to explore the galaxy with, what sort of moron would come **back to Earth**?". Also, if your twin is getting on a spacecraft to explore a planet around another star, why the hell aren't *you* going with them? Maybe there's something wrong with me, but the opportunity to explore beyond our own solar system was always more interesting than the idea of coming back to Earth.

If you *are* willing to go one-way to Mars, you'd also probably want to take a few things from Earth with you to make living like an underground mole-person a little more tolerable. Photos of family is a regular choice for astronauts heading to the International Space Station. Astronauts are usually allowed to take something about the size of a lunch box that can't weigh more than one kilogram, but maybe those going to Mars for good would want something a little more substantial. Music? Video games? Soft toys? What physical possession is so important that you'd pack it on a one-way trip to Mars? If you're *truly* starting a new life on Mars, do you pack anything at all? Personally, I've always said I'd take a ukulele, but that's mainly because I like the idea of my crewmates being trapped in a capsule on Mars with me as I torment them with a four-stringed nightmare machine for the rest of their lives...either that, or they make me practice in the airlock.

While they might be nice to have once you're living on Mars, small sentimental keepsakes like photos (and ukuleles) may play a critical role in the crew's mental health during the most psychologically and existentially challenging part of any human mission to Mars: the months of travel through interplanetary space when Earth has faded into the distance but Mars is yet to loom large in the window. When you're inside a spacecraft hurtling through the velvety blackness for months without any frame of reference, having something to emotionally and physically hold onto will be critical to the mental health of many on the crew; as that loss of planetary perspective is expected to be the biggest challenge for anyone who makes the journey.

Controlled Breakaway

Astronauts will need to metaphorically pass through the belly of the whale on their way to Mars, being truly disconnected from all that they've known before they begin the trials of life on the red planet. This is obviously where psychologists start freaking out about Breakaway Syndrome again, thinking that anyone who can't see Earth is going to experience a personal transformation with a rather concerning, "When you stare into the abyss, the abyss also stares into you"-Nietzsche kind of vibe to it. But around the same time Nietzsche was giving everyone an existential crisis in the late 1800's, Konstantin Tsiolkovsky was pioneering astronautics, developing the rocket equation, designing the first spaceships, and dropping little truth bombs like "The Earth is the cradle of the mind, but one cannot eternally live in a cradle". Personally I think 'cradle" is a bit brutal - I prefer to think that going to the Moon was a bit like going to a friend's for sleep over, and by colonising Mars we're finally moving out of our parents basement. Are we really ready to move out of home? I think *some* of us are. A small percentage of humanity has the sense of purpose and the 'right kind of crazy' needed to cross the threshold of our known experiences and take our next 'giant leap' to become a dual-planet species. Not *everyone* has to come to Mars, but those of us who look eagerly out into the cosmos will ultimately thrive compared to those that see Earth from above and long to return.

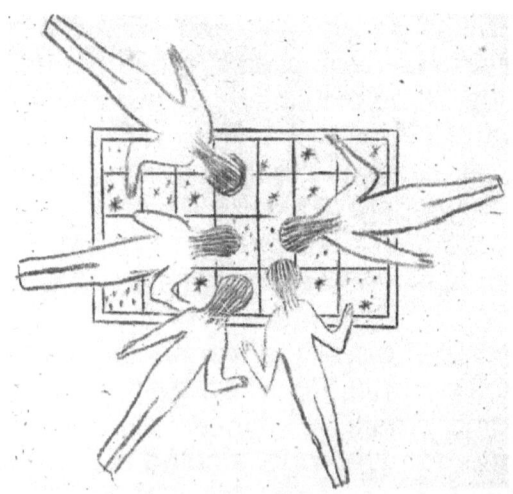

From Tsiolkovsky's 1933 paper "Album of Space Travel"
[Credit: Public Domain]

While Edgar Mitchell may have seen Earth from the Moon and subsequently raged against the small perspective of those who fight over petty Earth politics, he also experienced a separate, and incredibly vivid, spiritual experience on the way back from the Moon. To distribute the heat of the Sun evenly over the command and lunar modules during the three day flight back to Earth, the Apollo spacecraft were placed into a gentle rotation with respect to the Sun, commonly known as the "Barbecue Mode". With a little bit of spare time on his hands, Edgar Mitchell looked out of the Command Module's window to see the Earth, Sun, and Moon all against the backdrop of stars every 90 seconds. Mitchell was overcome by an overwhelming sense of interconnection to the universe. As a tiny bundle of carbon-based compounds inside an aluminum spacecraft, he was an inescapable and conscious link between the breathtaking beauty of our home planet, its moon, the sun, and the greater universe.

When Mae Jemison became the first African-American woman in space, she expressed a similar sense of connection: "My response when I went into space is that I was connected to everything. I felt much more connected to everything else in the universe, and sometimes on Earth I felt much more separate from the rest of the universe. I felt like I had as much right to be in space or in this universe as any speck of stardust. I was as eternal as that."

Through that extreme sense of connectedness and higher perspective, space amplifies who we are, while forcing us to acknowledge how small, insignificant, and unique we are as a species. While not everyone is going to want to move to Mars tomorrow, the increasing focus of sustainable, innovative technologies designed to improve life for all and to pursue understanding of our universe forces us to ask: what role can each of us play in supporting the next giant leap for humanity? Only twelve men walked on the moon during the Apollo program, yet over 400,000 worked for NASA to make the Apollo program a reality. What role would you be willing to play in the colonisation of Mars? Would you be someone like Michael Collins: vital to the success in the mission, but removed from the limelight by not being the one to take those historic first steps? Would you be willing to be in mission control instead of Mars? Or would you design and develop the plant-growing habitats? What about the countless supporters who would tune in each night, knowing they may not go to Mars themselves, but support those who do by watching them, sharing in the story of our species colonising another planet, and recognising how monumentally important it is for our species to have humans on another world- and being grateful to bear witness to it?

Trying to colonise Mars is a terrifyingly ambitious goal, but it may be the thing that finally sees us shift away from fighting for the resources of a single planet, and dividing ourselves along arbitrary lines. Instead it may help us recognise ourselves as a singular species capable of exploring this solar system and beyond. Even if specific plans to colonise Mars don't succeed, each attempt helps generate a critical mass around our hopes and dreams of exploring beyond the planet our species started on. In the words of Mars One ambassador and Nobel Laureate Gerard T'Hooft: "Maybe this project is more about having something to aim for". The technology to get humans to Mars and keep them alive there exists, but the risks will never be completely removed. What Mars really needs is a little courage from all of us. Courage to commit to something far bigger than any individual, courage to commit to something that will forever change the way we see ourselves as a species without really knowing how it will all turn out, and courage to leave behind our nationalism, our prejudice, and our egos when we do. If we can work together as a species to do all of that, then we'll be well on our way to becoming Martian.

Physicist, Explosives Engineer, Soldier, Comedian, Astronaut Candidate – one thing Josh Richards can never be accused of is being boring. After a decade of picking up booby traps with the Australian Army, slogging through mud with British Commandos, being science adviser to the richest artist in the world, and performing comedy wearing a giant koala suit to confused audiences around the world, Josh found his true calling in September 2012 when he discovered the Mars One project.

Selected from over 200,000 initial applicants, Josh is currently one of 100 astronaut candidates short-listed for a one-way mission to Mars in 2031. With a natural talent for explaining complicated science through comedy, he's a passionate and highly-visible ambassador for Science, Technology, Engineering, Arts and Math (STEAM) education. Besides regularly writing articles on space science, engineering, psychology and culture, his science communication events and school programs inspire people of all ages to engage with science, and in doing so, discover the sky is not the limit.

www.joshrichards.space
www.patreon.com/joshrichards

www.ingramcontent.com/pod-product-compliance
Lightning Source LLC
Chambersburg PA
CBHW032040290426
44110CB00012B/890